4/27/87

B.L. van der Waerden   A History of Algebra

B.L. van der Waerden

A History of Algebra

From al-Khwārizmī to Emmy Noether

With 28 Figures

Springer-Verlag
Berlin Heidelberg NewYork Tokyo

Prof. Dr. Bartel Leenert van der Waerden
Mathematisches Institut der Universität Zürich

Mathematics Subject Classification (1980):
01-XX, 10-XX, 12-XX, 14-XX, 16-XX, 17-XX, 20-XX, 22-XX

ISBN 3-540-13610-X Springer-Verlag Berlin Heidelberg New York Tokyo
ISBN 0-387-13610-X Springer-Verlag New York Heidelberg Berlin Tokyo

Library of Congress Cataloging in Publication Data
Waerden, B.L. van der (Bartel Leenert), 1903–
A history of algebra.
Bibliography: p. Includes index.
1. Algebra – History. I. Title.
QA151.W343 1985 512'.009 85-2820
ISBN 0-387-13610-X (U.S.)

Typesetting, printing and binding: Universitätsdruckerei H. Stürtz AG, D-8700 Würzburg
2141/3140-543210

Preface

The present volume is self-contained, but it is a part of a larger project. In an earlier volume, entitled "Geometry and Algebra in Ancient Civilizations", I have presented my view on the prehistory of algebra and geometry from the neolithical period to Brahmagupta (7th century A.D.). The present volume deals with the historical period from al-Khwārizmī, the earliest author of an "Algebra", to Emmy Noether.

In this book I shall restrict myself to three subjects, namely Part One: Algebraic Equations; Part Two: Groups; Part Three: Algebras.

My warmest thanks go out to all those who helped me by reading parts of the manuscript and suggesting essential improvements: Wyllis Bandler (Tallahassee, Florida), Robert Burn (Cambridge, England), Max Deuring[†] (Göttingen), Jean Dieudonné (Paris), Raffaella Franci (Siena), Hans Freudenthal (Utrecht), Thomas Hawkins (Boston), Erwin Neuenschwander (Zürich), Laura Toti Rigatelli (Siena), Warren Van Egmond (München).

In her usual careful way, Miss Annemarie Fellmann has typed the manuscript. She has also drawn the figures and helped me, together with Erwin Neuenschwander, in reading the proof sheets. Many thanks to both and to the editorial staff and the production department of the Springer-Verlag for their nice cooperation.

Zürich, March 1985   B.L. van der Waerden

Table of Contents

Part One. Algebraic Equations

Chapter 1. Three Muslimic Authors 3

   Part A. Al-Khwārizmī 3

     1. The Man and his Work 3
     2. Al-jabr and al-muqabala 4
     3. On Mensuration 5
     4. On the Jewish Calendar 7
     5. On Legacies 7
     6. The Solution of Quadratic Equations 7
     7. The Geography 9
     8. On Hindu Numerals 9
     9. The Astronomical Tables 9
    10. The "Sindhind" 10
    11. The "Method of the Persians" 11
    12. Al-Khwārizmī's Sources 13

   Part B. Tabit ben Qurra 15

   The Sabians 15
   The Life of Tabit ben Qurra 16
   On the Motion of the Eighth Sphere 17
   Geometrical Verification of the Solution of Quadratic Equations 18
   On Amicable Numbers 21

   Part C. Omar Khayyam 24

   The Algebra of Omar Khayyam 24
   Omar Khayyam on Ratios 29

Chapter 2. Algebra in Italy 32

   Part A. From Leonardo da Pisa to Luca Pacioli 32

   The Connection Between Trade and Civilization in Medieval Italy . . 32
   Life and Work of Fibonacci 33
   1. The "Liber Abbaci" 35
   2. The "Practica geometriae" 39
   3. The Book "Flos" 40
   4. The Letter to Theodorus 40
   5. The "Liber quadratorum" 40

Three Florentine Abbacists   42
1. Maestro Benedetto   42
2. Maestro Biaggio   43
3. Antonio Mazzinghi   44
Two Anonymous Manuscripts   45
Luca Pacioli   46

Part B. Master Dardi of Pisa   47

Part C. The Solution of Cubic and Biquadratic Equations   52

Scipione del Ferro   52
Tartaglia and Cardano   54
Lodovico Ferrari   56
Rafael Bombelli   59

Chapter 3. From Viète to Descartes   63
François Viète   63
Simon Stevin   68
Pierre de Fermat   69
René Descartes   72

Chapter 4. The Predecessors of Galois   76
Waring   76
Vandermonde   77
Lagrange   79
Malfatti   81
Ruffini   83
Cauchy   85
Abel   85

Chapter 5. Carl Friedrich Gauss   89
The Cyclotomic Equation   89
The "Fundamental Theorem"   94
The First Proof   95
The Second Proof   97
The Third Proof   99

Chapter 6. Evariste Galois 103
The Work of Galois 103
The Duel 104
The Memoir of 1831 105
Galois Fields 109
The Publication of Galois' Papers 112
Hermite, Puiseux, and Serret 112
Enrico Betti 114
The Second Posthumous Memoir of Galois 115

Chapter 7. Camille Jordan 117
Jordan's Traité 117
On Groups of Motions 118

On Congruences 121
Transitive and Primitive Groups of Substitutions 121
Series of Composition 121
Linear Substitutions 122
Jordan's Presentation of Galois Theory 124
Geometrical Applications 125
The 28 Double Tangents of a Plane Quartic 128
Application of Galois Theory to Transcendental Functions 131
On Solvable Groups 133

Part Two. Groups

Chapter 8. Early Group Theory 137
Part A. Groups of Substitutions 137
Early Theorems Concerning Subgroups of S_n 137
Mathieu 139
Sylow 139
Part B. Groups of Transformations 140
Non-Euclidean Geometry 141
Felix Klein and Sophus Lie 144
Felix Klein on Finite Groups of Fractional Linear Transformations . 145
Sophus Lie 146
Part C. Abstract Groups 147
Leonhard Euler 147
Carl Friedrich Gauss 148
Ernst Schering 149
Leopold Kronecker 149
Arthur Cayley 150
Walter von Dyck 152
Heinrich Weber 153
Part D. The Structure of Finite Groups 154
Otto Hölder 155
Finite Linear Groups 158
Chapter 9. Lie Groups and Lie Algebras 160
Part A. Lie Groups 160
Lie's Theory 160
Infinitesimal Transformations 162
Three Fundamental Theorems 163
Part B. Lie Algebras 165
Sophus Lie and Friedrich Engel 166
Wilhelm Killing 166
Élie Cartan 168
The Characteristic Roots 168

Semi-Simple Lie Groups 170
Weyl's Group (*S*) 172
Real Simple Lie Algebras 173

Part Three. Algebras

Chapter 10. The Discovery of Algebras 177
Complex Numbers 177
Hamilton's Discovery of Quaternions 179
The Leap into the Fourth Dimension 181
Octonions 183
Product Formulae for Sums of Squares 184
Geometrical Applications of Quaternions 185
The Arithmetic of Quaternions 186
Biquaternions 188
Full Matrix Algebras 189
Group Algebras 190
Grassmann's Calculus of Extensions 191
Clifford Algebras and Rotations in *n* Dimensions 192
Dirac's Theory of the Spinning Electron 194
Spinors in *n* Dimensions 195
Chevalley's Generalization 197
Generalized Quaternions 197
Crossed Products 199
Cyclic Algebras 200

Chapter 11. The Structure of Algebras 202
General Notions and Notations 202
Benjamin Peirce 203
Eduard Study 204
Gauss, Weierstrass and Dedekind 204
Georg Scheffers 205
Theodor Molien 206
Élie Cartan 208
Maclagan Wedderburn 210
Emil Artin 210
Emmy Noether and her School 211
Nathan Jacobson 211
Normal Simple Algebras 212
The Structure of Division Algebras 213
1. Division Algebras over $\mathbb{R}$ 214
2. Finite Skew Fields 214
3. Normal Simple Algebras over a *P*-adic Field 214
4. Division Algebras over an Algebraic Number Field 215

Chapter 12. Group Characters 218

   Part A. Characters of Abelian Groups 218

   Genera and Characters of Quadratic Forms 218

   Duality in Abelian Groups 221

   Part B. Characters of Finite Groups 223

   Dedekind's Introduction of the Group Determinant 223

   Frobenius on the Group Determinant 225

   Frobenius on Commuting Matrices 229

   The Letter of April 17, 1896 231

   The Letter of April 26 233

   Frobenius' Paper "Über Gruppencharaktere" 234

   The Proof of $e = f$ and the Factorization of the Group Determinant . 236

Chapter 13. Representations of Finite Groups and Algebras 237

   Heinrich Maschke 238

   Issai Schur 239

   Representations of the Symmetric Group 240

   The Representation of Groups by Projective Transformations 243

   Emmy Noether 244

   1. Group-Theoretical Foundations 244

   2. Non-Commutative Ideal Theory 245

   3. Modules and Representations 246

   4. Representations of Groups and Algebras 248

Chapter 14. Representations of Lie Groups and Lie Algebras 252

   Cartan's Theory 252

   The Global Method 253

   The Infinitesimal Method 254

   Hermann Weyl 257

   John von Neumann 261

Index 265

Part One
Algebraic Equations

Chapter 1
Three Muslimic Authors

It is beyond my competence to write a history of algebra in the Muslimic countries. Every year new publications on the subject appear. I guess the time has not yet come for a comprehensive history of Muslimic mathematics. Therefore I shall restrict myself to three most interesting authors, whose main works are available in modern translations, namely
A. Al-Khwārizmī,
B. Tābit ben Qurra,
C. Omar Khayyām.

Part A
Al-Khwārizmī

If we want to form an opinion on the scientific value and the sources of the work of al-Khwārizmī, we have to consider not only his treatise on Algebra, but also his other mathematical, astronomical, and calendaric work. The present section will be divided into twelve subsections.

1. The Man and his Work

An excellent account of the life and work of Muḥammad ben Mūsā al-Khwārizmī has been given by G.J. Toomer in Volume VII of the Dictionary of Scientific Biography, pages 358–365. From this account I quote:

Only a few details of al-Kwārizmī's life can be gleaned from the brief notices in Islamic bibliographical works and occasional remarks by Islamic historians and geographers. The epithet "al-Khwārizmī" would normally indicate that he came from Khwārizm (Khorezm, corresponding to the modern Khiva and the district surrounding it, south of the Aral Sea in central Asia). But the historian al-Ṭabarī gives him the additional epithet "al-Quṭrubbullī", indicating that he came from Quṭrubull, a district between the Tigris and Euphrates not far from Baghdad, so perhaps his ancestors, rather than he himself, came from Khwārizm; this interpretation is confirmed by some sources which state that his "stock" (aṣi) was from Khwārizm....

Under the Caliph al-Ma'mūn (reigned 813–833) al-Khwārizmī became a member of the "House of Wisdom" (Dār al-Ḥikma), a kind of academy of scientists set up at Baghdad, probably by Caliph Harūn al-Rashīd, but owing its preeminence to the interest of al-Ma'mūn, a great patron of learning and scientific investigation. It was for al-Ma'mūn that al-Khwārizmī composed his astronomical treatise, and his *Algebra* also is dedicated to that ruler.

From now on I shall omit all bars and dots. This simplifies the printing, and it will not give rise to any misunderstanding.

2. Al-jabr and al-muqabala

The biographer Haji Khalfa states in his biographical lexicon (ed. Flügel, Vol. 5, p. 67) that al-Khwarizmi was the first Islamic author to write "on the solution of problems by *al-jabr* and *al-muqabala*". What do these two expressions mean?

The usual meaning of *jabr* in mathematical treatises is: adding equal terms to both sides of an equation in order to eliminate negative terms. Another, less frequent meaning is: multiplying both sides of an equation by one and the same number in order to eliminate fractions. See George A. Saliba: The Meaning of al-jabr wa'l muqābalah, Centaurus 17, p. 189–204 (1973).

The usual meaning of *muqabala* is: reduction of positive terms by subtracting equal amounts from both sides of an equation. But al-Karaji also uses the word in the sense: to equate. The literal meaning of the word is: comparing, posing opposite.

The combination of the two words: *al-jabr wal-muqabala* is sometimes used in a more general sense: performing algebraic operations. It can also just mean: The science of algebra.

Let me give some examples of the use of these words in the work of al-Khwarizmi. On page 35 of Rosen's translation of the "Algebra of Mohammed ben Musa", the following problem is posed:

I have divided ten into two portions. I have multiplied the one of the two portions by the other. After this I have multiplied one of the two by itself, and the product of the multiplication by itself is four times as much as that of one of the portions by the other.

Al-Khwarizmi now calls one of the portions "thing" *(shay)* and the other "ten minus thing". Multipliying the two, he obtains in the translation of Rosen "ten things minus a square".

For the square of the unknown "thing" the author uses the word *mal*, which means something like "wealth" or "property". He finally obtains the equation

"A square, which is equal to forthy things minus four squares".

In modern notation, we may write this equation as

$$x^2 = 40x - 4x^2.$$

Next the author uses the operation *al-jabr*, adding $4x^2$ to both sides, thus obtaining

$$5x^2 = 40x$$

or

$$x^2 = 8x$$

from which he obtains $x = 8$.

Just so, on page 40, al-Khwarizmi has the equation

$$50 + x^2 = 29 + 10x$$

which is reduced by *al-muqabala* to

$$21 + x^2 = 10x.$$

In the introduction to his treatise the author states that the Imam al-Mamun

"has encouraged me to compose a short work on calculating by Completion and Reduction, confining it to what is easiest and most useful in arithmetic, such as men constantly require in cases of inheritance, legacies, partition, lawsuits, and trade, and in all their dealings with another, or where the measuring of lands, the digging of channels, geometrical computation, and other objects of various sorts and kinds are concerned…".

The full title of the treatise is "The Compendious Book on Calculation by *al-jabr* and *al-muqabala*". The treatise consists of three parts.

In the first part, al-Khwarizmi explains the solution of six types, to which all linear and quadratic equations can be reduced:

(1) $ax^2 = bx$

(2) $ax^2 = b$

(3) $ax = b$

(4) $ax^2 + bx = c$

(5) $ax^2 + c = bx$

(6) $ax^2 = bx + c,$

where a, b, and c are given positive numbers.

Al-Khwarizmi gives rules for solving these equations, he presents demonstrations of the rules, and he illustrates them by worked examples. We shall discuss his demonstrations presently.

3. On Mensuration

The second chapter of the "Algebra" is concerned with mensuration. Since Rosen's translation was deemed unsatisfactory, Solomon Gandz published the Arabic text together with a new English translation in his treatise "The Mishnat ha-Middot and the Geometry of Muhammed ibn Musa Al-Khowarizmi", Quellen and Studien zur Geschichte der Mathematik A2 (Springer-Verlag 1932).

The chapter consists mainly of rules for computing areas and volumes. For instance, the area of a circle is found by multiplying half of the diameter by half of the circumference. For finding the circumference, three rules are presented. If the diameter is d and the periphery p, the three rules are

(7)
$$p = 3\tfrac{1}{7}d,$$

(8)
$$p = \sqrt{10d^2},$$

(9)
$$p = \tfrac{62832}{20000}d.$$

Note that the rule (7) is due to Archimedes, who proved that p is less than 3 1/7 times d and more than 3 10/71 times d. The same rule (7) is also given by Heron of Alexandria in his "Metrica", and in the Hebrew treatise "Mishnat ha-Middot", edited and translated by Solomon Gandz.

The rule (8) is also found in Chapter XII of the Brahmasphutasiddhanta of Brahmagupta. See H.T. Colebrooke: Algebra with Arithmetic from the Sanskrit of Brahmegupta and Bhascara (London 1817, reprinted 1973 by Martin Sändig, Wiesbaden), p. 308–309.

Most remarkable is the rule (9), which is equivalent to the very accurate estimate

(10)
$$\pi \sim 3.1416.$$

Al-Khwarizmi ascribes the rule (9) to "the astronomers", and indeed the same rule is found in the *Aryabhatiya* of the Hindu astronomer Aryabhata (early sixth century AD). Verse II 28 of the Aryabhatiya reads:

Add 4 to 100, multiply by 8, and add 62000. The result is approximately the circumference of a circle of which the diameter is 20000 (see W.E. Clark: The Aryabhatiya of Aryabhata, p. 28).

In the last chapter of my book "Geometry and Algebra in Ancient Civilizations" (Springer-Verlag 1983) I have shown that the estimate (10) was also known to the Chinese geometer Liu Hui (third century AD). This estimate may well be due to Apollonios of Perge (see p. 196–199 and p. 207–213 of my book).

Al-Khwarizmi states that in every rectangular triangle the two short sides, each multiplied by itself and the products added together, equal the product of the long side multiplied by itself. Thus, if a, b, c are the sides, we have

$$a^2 + b^2 = c^2.$$

The proof given in the text is valid only in the equilateral case ($a=b$). From this fact we may safely conclude that al-Khwarizmi's main source is not a classical Greek treatise like the "Elements" of Euclid. Aristide Marre, who published a French translation of al-Khwarizmi's chapter on mensuration in Annali di matematica 7 (1866), noted the insufficiency of the proof and added that the author would never have been admitted to the Platonic academy!

An ancient Hebrew treatise exists which is closely connected, in contents and terminology, with Khwarizmi's chapter on Mensuration. The treatise is entitled "Mishnat ha-Middot". It was published, with an English translation and excellent commentary, by Solomon Gandz in Quellen und Studien zur Geschichte der Mathematik A2. By his arguments, Gandz has convinced me that the author of the treatise was *Rabbi Nehemiah*, who lived about AD 150.

The author knew how to compute the periphery of a circle as $3\frac{1}{7}d$. For the area of a circle segment, he presents the same formula as Heron of Alexandria:

$$A = (c+h) \cdot \frac{1}{2} h + \frac{1}{14}\left(\frac{c}{2}\right)^2$$

in which c is the chord and h the height of the segment. This formula is not in al-Khwarizmi's chapter on mensuration, but for the rest there are so many similarities between this chapter and the Mishnat ha-Middot, that one is forced to assume either a direct dependence, as Gandz does, or at least a common source. It is also possible, as Gandz supposes, that al-Khwarizmi used a Persian or Syrian translation of the Mishnat ha-Middot.

4. On the Jewish Calendar

No matter whether one does or does not accept the conclusion of Gandz that al-Khwarizmi's geometry was "verbally taken from the Mishnat ha-Middot", in any case al-Khwarizmi was acquainted with Jewish traditions, for he has written a treatise on the Jewish calendar. This treatise describes the Jewish 19-year cycle and the rules for determining on what weekday the month Tishri begins. It also calculates the interval between the Jewish "era of the creation of Abraham" and the Seleucid era, and it gives rules for determining the mean longitudes of sun and moon. See E.S. Kennedy: Al-Khwārizmī on the Jewish Calendar, Scripta mathematica 27, p. 55–59 (1964).

5. On Legacies

The third and largest part of the Algebra of al-Khwarizmi (p. 86–174 in Rosen's translation) deals with legacies. It consists entirely of problems with solutions. The solutions involve only simple arithmetic or linear equations, but they require considerable understanding of the Islamic law of inheritance. See Solomon Gandz: The Algebra of Inheritance, Osiris 5, p. 319–391 (1938).

6. The Solution of Quadratic Equations

I shall now discuss in somewhat greater detail al-Khwarizmi's solution of the three types of mixed quadratic equations. In al-Khwarizmi's own terminology, the first type reads:

Roots and Squares equal to numbers.

For instance: one square and ten roots of the same amount to thirty-nine dirhems; that is to say, what must be the square which, when increased by ten of its own roots, amounts to thirty-nine?

The solution is: you halve the number of roots, which in the present instance yields five. This you multiply by itself; the product is twenty-five. Add this to thirty-nine; the sum is sixty-four.

Now take the root of this, which is eight, and subtract from it half the number of the roots, which is four. The remainder is three. This is the root of the square you thought for; the square itself is nine.

In modern notation, the equation is

$$x^2 + 10x = 39,$$

which can be transformed into

$$(x+5)^2 = 39 + 25 = 64$$
$$x + 5 = \sqrt{64} \quad = 8$$
$$x = 8 - 5 \quad = 3.$$

Next, al-Khwarizmi presents a demonstration. He draws a square AB, the side of which is the desired root x. On the four sides he constructs rectangles, each having 1/4 of 10, or 2 1/2, as their breadth (see Fig. 1). Now the square

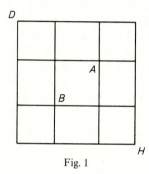

Fig. 1

together with the four rectangles is equal to 39. In order to complete the square DH, we must add four times the square of 2 1/2, that is, 25, says al-Khwarizmi. So the area of the large square is 64, and its side 8. Hence the side of the original square is

$$8 - 5 = 3.$$

Al-Khwarizmi next presents another, simpler proof, in which rectangles of breadth 5 are constructed only on two of the sides of the square AB (see Fig. 2). The result is, of course, the same.

Once more, we see that al-Khwarizmi's source is not Euclid, for his first proof is definitely more complicated than Euclid's proof of proposition II 4,

Fig. 2

which says that the square on a line segment $a+b$ is equal to the sum of the squares on a and b and twice the rectangle ab. The second proof of al-Khwarizmi is similar to that of Euclid.

I think this suffices to give the reader an idea of the style of al-Khwārizmī's treatise on *al-jabr* and *al-muqabala*. His treatment of the other types of mixed quadratic equations is quite similar to that of the first type. The other types are:

"Squares and numbers equal to roots",

"Roots and numbers equal to squares".

In each case, the solutions agree with those we learn at school, restricted to positive solutions.

7. The Geography

Besides the Algebra and the treatise on the Jewish calendar, one more treatise is extant in Arabic, namely the *Geography* ("Book of the Form of the Earth"). It consists almost entirely of lists of longitudes and latitudes of cities and localities. The work is a revision of Ptolemy's "Geography". Most probably it was based on a world map made by a commission of learned men (possibly including al-Khwarizmi himself) on the order of Caliph al-Mamun. For more details see Toomer's article al-Khwārizmī in the Dictionary of Scientific Biography VII, p. 361 and 365.

8. On Hindu Numerals

A treatise of al-Khwarizmi on Hindu numerals is extant only in a Latin translation, which was published first by B. Boncompagni under the title "Algoritmi de numero indorum" (Rome, 1857) and next by Kurt Vogel under the title "Mohammed ibn Musa Alchwarizmi's Algorithmus" (Aalen 1963), with a facsimile of the unique manuscript.

9. The Astronomical Tables

Al-Khwarizmi's set of astronomical tables is available only in a Latin translation of a revised version due to Maslama al-Majriti, who lived in Cordova about AD 1000. This version differed from the original text of Khwarizmi in several respects. First, the epoch of the original tables was the era Yazdigerd (16 June 632), whereas al-Majriti used the era Hijra (14 July 622). Also, al-Khwarizmi's table of Sines was based on the radius $R=150$, whereas the extant tables have $R=60$.

The tables have been published, with a German translation and commentary, by Heinrich Suter in Kongelige Dansk Vidensk. Selsk. Hist.-fil. Skrifter III, 1 (Copenhagen 1914). In the same Skrifter IV, 2 (Copenhagen 1962) Otto Neugebauer published an English translation of the introductory chapter and a new, valuable commentary. For additions and corrections to this commentary see C.G. Toomer's review in Centaurus 10, p. 203–212 (1964–65).

If one studies E.S. Kennedy's "Survey of Islamic Astronomical tables", Transactions American Philos. Soc. 46, p. 122–177 (1956), one sees that there are two types of *zijes*, i.e. astronomical table sets: Ptolemaic and Non-Ptolemaic. The Ptolemaic tables are based on Ptolemy's "Almagest" or on his "Handy Tables". The non-Ptolemaic *zijes*, of which al-Khwarizmi's table set is the only extant example, are based on Persian or on Hindu tables or on both. The non-Ptolemaic tables are less accurate, but more convenient than the Ptolemaic ones. This, I think, is the reason why Khwarizmi's tables remained popular long after the better (Ptolemaic) tables were available.

Ibn al-Qifti says in his biography of al-Fazari about al-Khwarizmi:

> He used in his tables the mean motions of the *Sindhind*, but he deviated from it in the equations (of the planets) and in the obliquity (of the ecliptic). He fixed the equations according to the method of the Persians, and the declination of the sun according to the method of Ptolemy.

What does this mean?

Let me begin with the last statement. In the *zij* of al-Khwarizmi there is a table for finding the declination of the sun (Suter's edition, p. 132–136, last column but one). This table is based on the value 23° 51′ of the obliquity of the ecliptic, and it agrees with a table in Ptolemy's "Handy Tables". So here al-Qifti is certainly right: the author of the tables determined the declination of the sun "according to the method of Ptolemy".

Concerning the "equations" of the planets, i.e. the corrections to be added to the mean longitudes, we may note that the maximum values of these corrections in the tables of al-Khwarizmi agree with those adopted in the Persian table set "*Zij-i Shah*". For this table set see E.S. Kennedy: The Sasanian Astronomical Handbook *Zīj-i Shāh*, Journal of the American Oriental Society 78, p. 246–262 (1958). Obviously, when al-Qifti speaks of "the Persians", he has the *Zij-i Shah* in mind, which was still extant in the time of al-Biruni and Ibn al-Qifti.

Thus we may conclude that one of the sources of al-Khwarizmi was the Persian table set "*Zij-i Shah*".

10. The "Sindhind"

I shall now discuss Ibn al-Qifti's first statement: "He used in his tables the mean motions of the Sindhind." The word Sindhind is a corruption of the Sanskrit *Siddhanta*, which is the usual designation of an astronomical textbook. In fact, the mean motions in the tables of al-Khwarizmi are derived from those in the "corrected Brahmasiddhanta" (Brahmasphutasiddhanta) of Brahmagupta. This was proved for the mean longitudes by J.J. Burckhardt, Vierteljahresschrift Naturf. Ges. Zürich 106, p. 213–231 (1961), and for the apogees and nodes by G.J. Toomer in his review of Neugebauer's commentary to al-Khwarizmi's tables (Centaurus 10, p. 207).

Soon after AD 770, a Sanskrit astronomical work called by the Arabs *Sindhind* was brought to the court of Caliph al-Mansur at Baghdad by a man called Kankah (or Mankah?), a member of a political mission from India. This work was translated into Arabic. Based on this translation, Yaqub ben Tariq, who is reported to have been at the court of al-Mansur together with Kankah, composed a table set, which was called *Zij al-Sindhind*. According to the Fihrist of el-Nadim (ed. Flügel, Vol. 1, p. 274) the table set of al-Khwarizmi was also called *Zij al-Sindhind*. It seems that al-Khwarizmi's *Zij* was a revision of an earlier table set based on the *Sindhind*, a revision into which some elements and methods from the *Zij-i Shah* and from Ptolemy's "Handy Tables" were incorporated.

11. The "Method of the Persians"

As we have seen, Ibn al-Qifti says that al-Khwarizmi "fixed the equations according to the Method of the Persians". What was this method?

I shall use the terminology and some notations of E.S. Kennedy's classical "Survey of Islamic Astronomical Tables" (Trans. Amer. Philos. Soc. 46). On pages 148–151 of this survey Kennedy presents an abstract of the tables of al-Khwarizmi, in which al-Khwarizmi's method of finding the true longitudes of the planets is explained.

Let $\bar{\lambda}$ be the mean longitude of any planet. Its true longitude is calculated by the formula

$$\lambda = \bar{\lambda} + e_1 + e_2,$$

where e_1 is the "equation of the centre" and e_2 the "equation of the anomaly". For the sun and the moon we have only one equation e_1 due to the eccentricity of the orbit. In al-Khwarizmi's tables for the sun and the moon, the function $e_1(x)$ is tabulated according to the formula

(12) $$e_1(x) = (\max e_1) \cdot \sin x$$

in which x is the distance of the mean sun or moon from the apogeum of the eccentric orbit:

(13) $$x = \bar{\lambda} - \lambda_{ap}.$$

For the other planets, the calculation is more complicated. One first calculates a preliminary value of the correction e_2, calculated by plane trigonometry from the triangle *EPM* in Fig. 3. In this drawing, the planet is supposed to be carried by an epicycle, which is in turn carried by a concentric circle. The angle e_2 can be tabulated as a function of the angle y (see H. Suter, Tafeln des Muhammed ibn Musa Al-Khwārizmī, pages 136–167, Column 3).

But, says Kennedy, "the inventor of the theory apparently realized that the two equations are not independent". We are required to halve the equation $e_2(y)$ and to add it to x, thus obtaining

(14) $$x' = x + 1/2\, e_2(y).$$

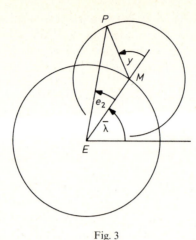

Fig. 3

This x' is used to calculate the correction e_1:

(15) $e_1(x')=(\max e_1)\cdot\sin x'$

which is subtracted from y, thus obtaining

(16) $y'=y-e_1(x')$.

Now the longitude λ can be calculated as

(17) $\lambda=\bar{\lambda}+e_1(x')+e_2(y')$.

So one has to use the table for e_1 twice, first to find $e_1(x)$ and next $e_1(x')$, and the table for e_2 once to find $e_2(y')$. For the rest, one has to perform only simple additions and subtractions. The procedure is simpler, but less accurate than Ptolemy's method.

As we have seen, al-Khwarizmi used in his tables the "Era Yazdigerd". So we may safely conclude that he learnt the "Method of the Persians" from the latest version of the *Zij-i Shah*, which was composed under the last Sasanid king Yazdigerd III (632–651). See for the history of this version pages 4–5 of a joint paper of J.J. Burckhardt and myself: Das astronomische System der persischen Tafeln, Centaurus 13, p. 1–28 (1968).

In earlier, predominantly Hindu texts we find a related, but slightly more complicated method, which we have called "Method of the Indians". It is based on the formulae

(14) $x'=x+1/2\,e_2(y)$

(15) $e_1(x')=(\max e_1)\cdot\sin x'$

(18) $x''=x'+1/2\,e_1(x')$

(19) $$e_1(x'') = (\max e_1) \cdot \sin x''$$

(20) $$y' = y - e_1(x'')$$

(21) $$\lambda = \bar{\lambda} + e_1(x'') + e_2(y').$$

This method was used by Aryabhata (Aryabhatiya, verses 22–24), by Brahmagupta (Brahmasphutasiddhanta II, 34–38), and by other Hindu astronomers. The difference as compared with the Persian method is that the table for $e_1(x)$ is used twice: once with the argument x' and once with the argument x''. The difference is only small, for $1/2\, e_1(x')$ is in most cases small, so that x'' defined by (18) does not differ much from x'.

In my paper "Ausgleichspunkt, Methode der Perser und indische Planetenrechnung", Archive for History of Exact Sciences 1, p. 107–121 (1961), I have shown that the "Method of the Indians" can be explained as a reasonable approximation, if we suppose that a Greek author before Ptolemy, possibly Apollonios of Perge, started with the model of an epicycle carried by an eccentric circle. I suppose that this author assumed an "equant point" as in Ptolemy's Almagest, such that the motion on the eccenter appeared uniform as seen from the equant point. He invented an approximation which enabled the user of the tables to use only one-entry tables and additions and subtractions. Ptolemy adopted the equant model, but he did not use the approximation. On the other hand, the Hindu authors adopted the simple method of calculation, probably without knowing that it was based on the assumption of an eccenter with equant point.

This seems to be the only hypothesis which explains Ptolemy's equant model, for which Ptolemy himself gives no justification whatever, as well as the very sophisticated "Method of the Indians", for which the Hindu authors give no justification either.

12. Al-Khwarizmi's Sources

We are now in a position to discuss the sources of al-Khwarizmi's work, in particular of his Algebra. Three theories have been proposed. He may have used classical Greek sources, or Hindu sources, or popular mathematical writings belonging to the Hellenistic and post-Hellenistic tradition.

As Toomer notes in his article in the Dictionary of Scientific Biography, both Greek and Hindu algebra had advanced well beyond the elementary stage of al-Khwarizmi's work, and none of the known works in either culture shows much resemblance in presentation to al-Khwarizmi's work. As we have seen, his proofs of the methods of solution of quadratic equations are quite different from the proofs we find in Euclid's Elements. Also, as Toomer notes, al-Khwarizmi's exposition is completely rhetorical, like Sanskrit algebraic works, and unlike the one surviving Greek algebraic treatise, that of Diophantos, which has already developed quite far towards symbolic representation.

I feel that Toomer is right: we may exclude the possibility that al-Khwarizmi's work was much influenced by classical Greek mathematics.

In favour of the Hindu hypothesis it may be argued that al-Khwarizmi did write a treatise on Hindu numerals, that two of his estimates for π are also found in Hindu sources, and that in Chapter 18 of the *Brahmasphutasid-dhanta* of Brahmagupta, verse 18, a general rule for the solution of a quadratic equation of type (4) is given. See for this rule H.T. Colebrooke: Algebra with Arithmetic and Mensuration from the Sanskrit of Brahmagupta and Bhascara, page 346.

In one case, in the section on Mensuration, al-Khwarizmi gives us a hint concerning his sources. After having mentioned the estimate $3 + 1/7$ for π, which is "generally followed in practical life, though it is not quite exact", he says:

The mathematicians, however, have two other rules for that. The one of them is: multiply the diameter with itself, then with ten, and then take the root of the product. The root gives the circumference.

The other rule is used by *the astronomers among them* (my italics), and reads: multiply the diameter with sixty-two thousand eight hundred and thirty-two and then divide it by twenty thousand. The quotient gives the circumference.

Note that Aryabhata writes his estimate of π in just the same form as

$$62\,832/20\,000.$$

We know already that al-Khwarizmi used Persian and Hindu sources in composing his astronomical tables. We may suppose that he derived his estimate of π from one of these sources.

After the Greek and the Hindu hypotheses, we may discuss a third hypothesis proposed by Hermann Hankel in his "Geschichte der Mathematik" (Leipzig 1874), p. 259–264, and supported by H. Wiedemann in his article "al-Khwārizmī" in the Encyclopaedia of Islam II, p. 912–913. These authors deny all Greek influence on al-Khwarizmi and assert the prevalence of a native, Syriac-Persian tradition.

In view of the close connection between the Hebrew treatise Mishnat ha-Middot and the geometry of al-Khwarizmi, I feel we should extend the notion "Syriac-Persian" to include Hebrew and other popular traditions as well. We have to admit the existence of a tradition of popular mathematics in Egypt and in the Near East in Hellenistic and post-Hellenistic times. Examples are the mathematical papyri from Egypt discussed on pages 164–170 and 173–177 of my "Geometry and Algebra in Ancient Civilizations", and the "Metrica" of Heron of Alexandria discussed on pages 181–188 of the same book.

The hypothesis of Hankel and Wiedemann was strongly supported by Solomon Gandz, the editor of the "Mishnat ha-Middot". I think I can do no better than quote the final section of his introduction to the Mensuration of al-Khwarizmi:

Al-Khowārizmī, the antagonist of Greek influence

At the university of Baghdad founded by al-Ma'mūn (813–33), the so-called Bayt al-Ḥikma, "the House of Wisdom", where al-Khowārizmī worked under the patronage of the Caliph, there and then flourished also an older contemporarara of al-Khowārizmī by the name of al-Ḥajjāj ibn Yūsuf ibn Maṭar. This man was the foremost protagonist of the Greek school working for the reception of Greek science by the Arabs. All his life was devoted to the work on Arabic translations from the Greek. Already under Harūn al-Rashīd (786–809) al-Ḥajjāj had brought out an Arabic translation of Euclid's Elements. When al-Ma'mūn became Caliph, al-Ḥajjāj tried to gain his favor by preparing a second improved edition of his Euclid translation. Later on (829–30) he translated the Almagest. Now al-Khowārizmī never mentions this colleague of his and never refers to his work. Euclid and his geometry, though available in a good translation by his colleague, is entirely ignored by him when he writes on geometry. On the contrary, in the preface to his Algebra al-Khowārizmī distinctly emphasizes his purpose of writing a popular treatise that, in contradiction to Greek theoretical mathematics, will serve the practical ends and needs of the people in their affaires of ineritance and legacies, in their law suits, in trade and commerce, in the surveying of lands and in the digging of canals. Hence, al-Khowārizmī appears to us not as a pupil of the Greeks but, quite to the contrary, as the antagonist of al-Hajjaj and the Greek school, as the representative of the native popular sciences. At the Academy of Baghdad al-Khowārizmī represented rather the reaction against the introduction of Greek mathematics. His Algebra impresses us as a protest rather against the Euclid translation and against the whole trend of the reception of the Greek sciences.

Part B
Tabit ben Qurra

The Sabians

The great scientist Tabit ben Qurra al Harrani (836–901) was a "Sabian" from Harran. What does this mean? In my explication I shall follow the two-volume standard work of D. Chwolson: "Die Ssabier und der Ssabismus" (St. Petersburg 1856, reprinted by Oriental Press, Amsterdam 1965).

According to Chwolson, we have to distinguish between two kinds of Sabians: the genuine or *Chaldaean Sabians* and the *pseudo-Sabians from Harran*, to which Tabit ben Qurra and al-Battani belonged.

The Chaldaean Sabians are mentioned in the Koran (II 59 and XXII 17) among the believers in God, who have sacred books and shall not be persecuted.

Who were these Sabians? According to Chwolson, they were identical with the *Mandaeans*, a gnostic sect living in Southern Mesopotamia near the moors and lakes of Basra. See D. Chwolson: Die Ssabier I, p. 100–143, and E.S. Drower: The Mandaeans of Iraq and Iran (1962).

The Sabians of Harran were quite different from the genuine Sabians mentioned in the Koran. The historian Mas'udi says that the Sabians "who have their homes in Wâsith and in Basrah in Iraq differ from the Sabians of Harran by their outer appearance" (see Chwolson: Die Ssabier II, p. 376). Also, their religion was different. For the Mandaeans in Southern Mesopotamia the seven planets and the twelve zodiacal signs were evil powers, but the Harranites built temples for the planetary gods (see Chwolson II, p. 1–52 and 366–379).

In the present chapter we are mainly concerned with the Harranite Sabians. Their way of life was in several respects similar to the "Pythagorean Life" as described by the Neo-Pythagorean Iamblichos (see my book "Die Pythagoreer", pages 319–320). For instance, the Sabians as well as the Pythagoreans were not allowed to eat beans.

These similarities were not accidental: the Sabians were fully aware that they continued the tradition of the Pythagoreans. We happen to know that Tabit ben Qurra translated two Neo-Pythagorean writings from Greek into Arabic, namely the Arithmetical Introduction of Nikomachos of Gerasa, and a part of the commentary of Proklos to the Pythagorean Golden Verses. See Chwolson, Die Ssabier I, p. 559.

I shall now tell the story of how the Harranians came to call themselves "Sabians". The story is phantastic and difficult to believe, but if one studies the testimonies quoted in the book of Chwolson, one cannot but agree with his conclusions.

A Christian author named Abu Yusuf Yashu al-Qati'i, who lived at the end of the 9th century, wrote a book intending to reveal "the doctrines of the Harranians who are known in our time as Sabians". From this book we have an apparently verbal excerpt in the Fihrist of al-Nadim (see Chwolson II, p. 14–19). Abu Yusuf relates that the Caliph al-Mamun, on his campaign against the Byzantine emperor, came to Harran and asked the inhabitants

"Are you Christians?" "No."

"Are you Jews?" "No."

On his next question "Have you got a sacred book or a prophet?" he did not get a clear answer. So the Caliph said: Either you convert to Islam or to one of the religions admitted by the Koran, or you shall be killed when I return from my expedition.

Now a sheik from Harran, who was versed in Moslem law, gave them (against good payment) the advise: Call yourself Sabians, for this is the name of a religion recognized in the Koran. This they did, and from now on they were called Sabians.

The Life of Tabit ben Qurra

According to the Fihrist (see Chwolson, Die Ssabier I, p. 532 and 547), Tabit ben Qurra el-Harrani died in AD 901 and lived 77 solar years. This would imply that he was born AD 824, but the Fihrist says that he was born AD 836, and other sources give 826 as his birth year.

In his native town Harran he lived as a money changer, but his ideas about religion led to a conflict (see Chwolson I, p. 482–490). He was brought before the highest priest, who declared his doctrines heretical and prohibited his entrance to the temple. Chwolson thinks that Tabit's neo-Platonic philosophy was judged a heresy. He first revoked his opinions, but afterwards he stated them anew. He was excommunicated, and he left the city. It so happened that he met Mohammed ben Musa ben Sakir, one of the famous "sons of Musa": Mohammed, Ahmed, and Hasan, who were great collectors of books and great patrons of science (see H. Suter: Die Mathematiker und Astronomen der Araber, p. 20–21). This Mohammed ben Musa took Tabit to Baghdad, allowed him to live in his house, and introduced him to the Caliph. All this must have happened before AD 873, for in January 873 Mohammed ben Musa died (see Suter: Die Mathematiker and Astronomen, p. 20).

According to al-Nadim and el-Qifti (see Chwolson I, p. 483 and II, p. 532) Tabit succeeded in establishing at Baghdad a Sabian primate for the whole of Iraq. By this move, the situation of the Sabians was stabilized, and they were respected in the whole country.

Tabit was highly esteemed for his writings in medicine, philosophy, mathematics, astronomy, and astrology. He was also a most competent translator from Greek and Syriac into Arabic. He translated works of Euclid, Archimedes, Apollonios, Autolykos, Ptolemaios, Nikomachos, Proklos, and others (see Chwolson I, p. 553–560).

Barhebraeus reports in his Syrian chronicle that Tabit ben Qurra composed circa 150 works in Syriac. For his works on astronomy and mathematics see H. Suter: Die Mathematiker und Astronomen der Araber (1900), p. 34–38, and Nachtrag, p. 162–163. Here I shall restrict myself to three treatises: one on astronomy, one on algebra, and one on arithmetic.

On the Motion of the Eighth Sphere

Tabit has written a very interesting treatise, which is available only in a Latin translation, entitled "De motu octave spere". The Latin text was published by C.F. Carmody: "The Astronomical Works of Tabit b. Qurra" (Berkeley 1960), p. 84–113. An English translation with commentary was presented by O. Neugebauer in Proceedings of the Amer. Philos. Soc. 106, p. 291–299.

The "eighth sphere" of Tabit is the sphere of the fixed stars. Inside this sphere one has to imagine the seven spheres of the moon, the sun, and the five "star-planets".

In modern astronomy the fixed stars are assumed to be nearly at rest and the equinoxes to have a small *retrograde* motion with respect to the fixed stars: the "precession of the equinoxes". In Ptolemy's theory the equinoxes are fixed, and the stars are supposed to have a slow *forward* motion of 1 degree in 100 years.

Tabit noticed that this small amount is not confirmed by the observations. The motion of the stars with respect to the equinoxes has to be much larger, at least in the time after Ptolemy, if one accepts the very accurate observations made under the reign of al-Mamun. To explain this, Tabit assumed an oscillatory (periodic) motion of the sphere of the fixed stars, the so-called "trepidation".

Another phenomenon which Tabit wanted to explain is an alleged decrease of the obliquity of the ecliptic. The ancient Greeks had used a rough estimate of 24°, Ptolemy had used a slightly smaller estimate due to Eratosthenes, and the observers at Baghdad had found a still smaller obliquity, namely 23° 33'.

Tabit now constructed a model which would explain both phenomena: the alleged trepidation of the fixed stars with respect to the equinoxes, and the alleged decrease of the obliquity. He made the two opposite points "Beginning of Aries" and "Beginning of Libra" on the sphere of the fixed stars move slowly on small circles, whose centres are opposite points of a fixed sphere. For a detailed description of this model I may refer to the paper of Neugebauer just mentioned.

Tabit's treatise ends up with two small tables, from which the motion of the two variable points "Beginning of Aries" and "Beginning of Libra" can be computed.

Geometrical Verification of the Solution of Quadratic Equations

Tabit's short treatise on this subject, entitled "On the Verification of Problems of Algebra by Geometrical Proofs", is preserved in a single manuscript Aya Sophia 2457,3. It was published with a German translation and commentary by P. Luckey in 1941: Berichte über die Verhandlungen der sächs. Akad. Leipzig 93, p. 93–112. I shall now translate parts of Luckey's translation into English. Since the logic of the treatise is perfect, I see no danger in this procedure. The diagrams are not taken from the manuscript, but from Luckey's publication.

There are three fundamental forms (uṣūl, roots or elements), to which most problems of algebra can be reduced:

The first basic form is: Wealth (māl) and roots are equal to numbers. The way and method of solution by the sixth proposition of Euclid's second book is as I shall describe: We make (Fig. 4) the wealth equal to the square $abgd$, we make bh equal to the same multiple of the unit in which lines are measured as is in the given number of roots, and we complete the area dh. Since the wealth is $abgd$, the root is clearly ab, and in the domain of calculation and number it is equal to the product of ab and the unit, in which the lines are measured …. Now a number of these units equal to the given number of roots is in bh, hence the product of ab and bh is equal to the roots in the domain of calculation and number. But the product of ab and bh is the area dh, because ab is equal to bd. Hence the area dh is in this way equal to the roots of the problem. Hence the whole are gh is equal to the wealth together with the roots.

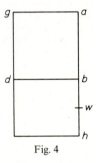

Fig. 4

Tabit's explanation is cumbersome, because he cannot equate an area or line segment with a number. He therefore introduces a unit of length, which I shall denote by e. If the given equation is

$$x^2 + mx = n,$$

in which x is an unknown number, while m and n are given numbers, he translates it into a geometrical equation

$$x^2 + mex = ne^2$$

in which x and e are line segments. He continues:

Now the wealth and the roots together are equal to a known number. So the area gh is known, and it is equal to the product of ah and ab, because ab is equal to ag. So the product of ha and ab is known, and the line bh is known, because its number of units is known. Thus everything is reduced to a well-known geometrical problem, namely: The line bh is known. To it a line ab is added, and the product of ha and ab is known.

Now in proposition 6 of book 2 of the Elements it is proved that, if the line bh is halved at the point w, the product of ha and ab together with the square of bw is equal to the square of aw. But the product of ha and ab is known, and the square of bw is known. So the square of aw is known, hence aw is known, and if the known bw is subtracted, ab results as known, and this is the root. And if we multiply it by itself, the square $abgh$, that is, the wealth, is known, which is what we wanted to prove.

Now comes the most interesting passage in the treatise:

This procedure agrees with the procedure of the people concerned with algebra in their solution of the problem. When they halve the number of roots, this is just so as when we take half of the line bh, and when they multiply it by itself, this is the same as when we take the square of the halved line bh. When they add to the result the (given) number, this is just as when we add the product of ha and ab, in order to obtain the square of the sum of ab and the halved line. Their taking the root of the result is like our saying: The sum of ab and the halved line is known as soon as its square is known.

The next sentence in the text is corrupt. The end of the sentence reads:

... to obtain the residue, just as we obtained ab. They multiplied (the residue) by itself, just as we determined the square of ab, that is, the wealth.

In the same way Tabit treats the second type of equation

$$x^2 + b = ax$$

or "wealth and number is equal to roots". He says:

The way and method of solution according to the second book of Euclid by means of proposition 5 is, as I describe it: We make (Fig. 5) the wealth into a square $abgd$ and we make ah equal to such a multiple of the unit in which lines are measured as is in the given number of roots. Obviously, ah is longer than ab, because the roots, which are in the domain of calculation the product of ga and ah, are larger than the wealth. We complete the area gh, and we prove, as before, that it is equal to the roots (that is, equal to the term ax) in the domain of calculation. And if bg, which is the wealth (that is, the term x^2) is subtracted from it, there remains dh equal to the (given) number. So dh is known, and it is equal to the product of ab and bh, and the line ah is known. So now the problem amounts to dividing a given line ah in b in such a way that the product of ab and bh is known.

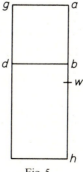

Fig. 5

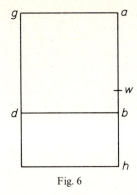

Fig. 6

Now in proposition 5 of the second book of Euclid it is proved that, if ah is halved at w, the product of ab and bh together with the square of bw is equal to the square of aw. But aw is known, and its square is known, and the product of ab and bh is known. So the square of bw is known as a remainder, hence bw is known, and if it is subtracted from aw (Fig. 5) or added to it (Fig. 6), ab results as known, and it is the root. And if we multiply it by itself, $abgd$ is known, and it is the wealth, and this is what we wanted to prove.

This procedure too agrees with the procedure of the algebra people *(ahl al-jabr)* in calculating the problem. For it allows in both ways the application of addition and of subtraction of the line wb.

I think it is not necessary to translate the third part of the text, in which the equation

"Number and Roots are equal to Wealth"

is solved by means of Euclid's proposition II 6, and the agreement with the algebraic solution is proved in the same way as in the other two cases.

In al-Khwarizmi's treatise, the science of algebra is denoted by the double expression "*al-jabr wal muqabala*". Tabit ben Qurra leaves out the second part and refers just to the "solution by al-jabr" as opposed to his own solution by geometry. The algebrists, to which al-Khwarizmi belongs, are called by Tabit "those concerned with algebra" *(ashāb aljabr)* or "the algebra people" *(ahl al-jabr)*. In the text, they are opposed to the geometers, to which Tabit himself belongs.

Tabit judges it necessary to explain in great detail that the algebraic solutions are in full accordance with Euclid's geometrical solution. From this, Luckey concludes that at least for some of his readers this connection between geometry and algebra was new, and he raises the question: Was it new for the "algebra people"? It seems to me that the answer must be "yes", for otherwise the whole treatise of Tabit would be superfluous.

As we have seen in the section on al-Khwarizmi, there were two opposite trends or parties among the mathematicians and astronomers at Baghdad. One of these trends was represented by al-Khwarizmi, who used Indian and Persian sources for his astronomical tables, and who wrote his Algebra, "confining it to what is easiest and most useful in arithmetics, such as men constantly require in cases of inheritance", and so on. On the other hand, we have "the Greek school working for the reception of Greek science by the Arabs", as Gandz puts it. To this Greek school belonged al-Hajjaj, who translated Euclid and Ptolemy, and Tabit ben Qurra.

On Amicable Numbers

Two natural numbers m and n are called *amicable*, if each is equal to the sum of the proper divisors of the other. For instance, the sum of the proper divisors of 284 is 220, and the sum of the proper divisors of 220 is 284. This pair of amicable numbers was known already to the ancient Pythagoreans (see e.g. my "Science Awakening" I, p. 98).

Tabit ben Qurra has written a "Book on the Determination of Amicable Numbers". He proved: If $p = 3 \cdot 2^{n-1} - 1$ and $q = 3 \cdot 2^n - 1$ and $r = 9 \cdot 2^{2n-1} - 1$ are prime, then

$$M = 2^n p q \quad \text{and} \quad N = 2^n r$$

are amicable numbers.

Tabit's book has been commented upon and partly translated by F. Woepcke: Notice sur une théorie ajoutée par Thābit ben Korra à l'arithmétique spéculative des Grecs, Journal asiatique (4) 20, p. 420–429 (1852).

Tabit's rule for obtaining amicable pairs was rediscovered by *Pierre de Fermat* and *René Descartes*. Besides the well known pair 220 and 284, Fermat found one more pair, namely

$$17296 = 2^4 \times 23 \times 47$$
$$18416 = 2^4 \times 1151$$

(Oeuvres II, p. 20–21). No doubt, he derived it by Tabit's rule for $n = 4$.

Descartes formulated Tabit's rule explicitly and presented a third example:

$$9363584 = 2^7 \times 191 \times 383$$
$$9437056 = 2^7 \times 73727$$

(René Descartes, Oeuvres II, p. 93–94 and p. 148).

Now the question arises: How did Tabit find his rule?

The well known pair 220 and 284 has a factorization of the form

$$2^2 p q \quad \text{and} \quad 2^2 r$$

in which p, q, and r are primes. So let us see whether we can find a pair

$$M = 2^n p q \quad \text{and} \quad N = 2^n r$$

such that M is the sum of the proper divisors of N and conversely.

I suppose that Tabit knew that the sum of all divisors of N (including N itself) is

$$(1 + 2 + \ldots + 2^n)(r + 1)$$

and that the sum of all divisors of M is

$$(1 + 2 + \ldots + 2^n)(pq + p + q + 1).$$

Now both sums are required to be equal to $M+N$. So we must have

(1) $$r=pq+p+q$$

and

(2) $$(2^{n+1}-1)(pq+p+q+1)=2^n pq+2^n r.$$

Substituting (1) into (2), one obtains a condition for p and q:

(3) $$(2^{n+1}-1)(pq+p+q+1)=2^n pq+2^n(pq+p+q).$$

It is easy, though a little clumsy, to formulate the derivation of (3) in the language of "rhetorical algebra" used by Tabit. By the operations *al-jabr* and *al-muqabala*, (3) can be simplified to

(4) $$2^n(p+q+2)=pq+p+q+1.$$

Putting $p+1=P$ and $q+1=Q$, (4) can further be simplified to

(5) $$2^n(P+Q)=PQ.$$

Adding 2^{2n} to both sides, and subtracting $2^n(P+Q)$, one obtains

$$2^{2n}=PQ-2^n P-2^n Q+2^{2n}$$

or

$$2^{2n}=(P-2^n)(Q-2^n).$$

The two factors on the right hand side are either both positive or both negative. If they were both negative, their product would be less than 2^{2n}, so they must be positive. Since their product is 2^{2n}, we must have, assuming $P<Q$,

$$P-2^n=2^{n-t}$$
$$Q-2^n=2^{n+t}.$$

The simplest choice of t is $t=1$, which leads to

$$P=2^n+2^{n-1}=3\times2^{n-1}$$
$$Q=2^n+2^{n+1}=6\times2^{n-1}$$

and thus to Tabit's solution

$$p=3\times2^{n-1}-1$$
$$q=3\times2^n-1$$
$$r=PQ-1=9\times2^{2n-1}-1.$$

Another possibility is $n=8$ and $t=7$, which leads to a pair of amicable numbers discovered by Legendre in 1830: $2^8 \times 257 \times 33023$ and $2^8 \times 8520191$.

In the preceding calculation, we have assumed two algebraic identities, namely

(6) $$(p+1)(q+1)=pq+p+q+1$$

(7) $$(P-2^n)(Q-2^n)=PQ-2^nP-2^nQ+2^{2n}$$

which are both easy to prove by the methods of Euclid's Book 2. So the derivation just given is well within the range of algebraic methods known to Tabit.

After Fermat and Descartes, Leonard Euler was the first to take up the problem of amicable numbers. Euler has written three papers on the subject, all entitled "De numeris amicabilibus". In the first paper of 1747 (Opera omnia, series prima, Vol. 2, p. 59–61) Euler presented a derivation of the rule of Tabit along the lines indicated here, and a list of 30 pairs of amicable numbers. In the second paper of 1750 (Opera omnia, same volume, p. 23–107) Euler gave a full exposition of his methods and presented an extended list of 62 pairs. In his exposition, he solved two problems:

Problem 1. To find two amicable numbers apq and ar such that p, q, and r are primes. Denoting by A the sum of all divisors of a and putting, as before, $P=p+1$ and $Q=q+1$, he finds an equation analoguous to our equation (5), namely

(8) $$a(P+Q)=(2a-A)PQ.$$

Putting

$$\frac{a}{2a-A}=\frac{b}{c}, \quad (b,c)=1$$

Euler obtains

$$(cP-b)(cQ-b)=b^2.$$

So, in order to find P and Q, one has to factorize b^2 into two different factors.

By similar methods, Euler solves *Problem 2:* To find two amicable numbers apq and ars.

In his third paper, published posthumously in Opera Omnia, series prima, Vol. 5, p. 353–365, Euler presents four more examples, all of the form

$$apq \quad \text{and} \quad ar.$$

For more information about amicable numbers see the survey of Edward B. Escott in Scripta Mathematica 12, p. 61–72 (1946).

Part C
Omar Khayyam

The Persian poet, philosopher, mathematician, and astronomer Omar ben Ibrahim al-Hayyam, usually called Omar Khayyam, lived in the second half of the eleventh century. His fame in the western world is mainly based on the very free translation of his nearly 600 short poems of four lines each (*Ruba'iyat*) by E. Fitzgerald (1859),

In 1074 Omar Khayyam was called to Isfahan, where a group of outstanding astronomers came together for the foundation of an observatory. "An enormous amount of money was spent for this purpose", says Ibn al Athir. See Aydin Sahili: The Observatory in Islam (Türk Kurumu Basimevi, Ankara 1960).

Here we shall mainly be concerned with Omar Khayyam's treatise "On the Proofs of the Problems of Algebra and Muqabala". My account will be based on the French translation of Franz Woepcke: L'algèbre d'Omar Alkhayyami (Paris 1851). An English translation was published in 1950 by H.J.J. Winter and W. Arafat in Journal R. Asiatic Soc. Bengal. 16, p. 27–77. For an edition of the text with a new French translation and commentary see Roshdi Rashed and Ahmed Djebbar: L'oeuvre algébrique d'Al-Khayyam, University of Aleppo 1981.

In the introduction to his "Algebra" Omar Khayyam explains that "The art of algebra" aims at the determination of *numerical* or *geometrical* unknown quantities. This distinction between *numbers* and *measurable magnitudes* is maintained throughout the treatise. The author mentions four kinds of measurable magnitudes: the *line*, the *surface*, the *solid*, and the *time*. He excludes magnitudes of more than three dimensions such as the "square-square" and the "quadrato-cube", which are used by some algebrists.

The Algebra of Omar Khayyam

The algebra of Omar Khayyam is mainly geometric. He first solves linear and quadratic equations by the geometrical methods explained in Euclid's Elements, and next he shows that cubic equations can be solved by means of intersections of conics.

Omar knows very well that earlier authors sometimes equated geometrical magnitudes with numbers. He avoids this logical inconsistency by a trick, introducing a unit of length. He writes:

Every time we shall say in this book "a number is equal to a rectangle", we shall understand by the "number" a rectangle of which one side is unity, and the other a line equal in measure to the given number, in such a way that each of the parts by which it is measured is equal to the side we have taken as unity.

In Fig. 7 I have denoted the unity of length by e, and the sides of the rectangle by x and y. The figure illustrates the equation $3 = xy$.

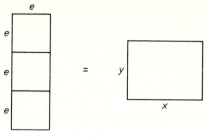

Fig. 7

Omar Khayyam first solves quadratic equations by the usual methods. Next he passes to cubic equations. Some of these, for instance,

(1) $$x^3 + ax^2 = bx$$

can be reduced to quadratic equations. The first type requiring conic sections is

"A number is equal to a cube"

or, in modern notation

(2) $$x^3 = N.$$

Omar first solves an auxiliary problem, namely
"To find two lines between two given lines such that the four lines form a continued proportion".

If the two given lines are called $AB = a$ and $BC = b$, the problem is, to find x and y such that

(3) $$a:x = x:y = y:b.$$

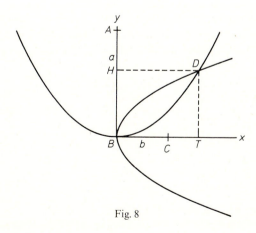

Fig. 8

Omar draws two perpendicular line segments BA and BC, and he constructs two parabolas, both having their summit at B. The first parabola has axis BC and "parameter" BC, the other has axis BA and "parameter" BA. In modern notation, the equations of the two conics are

(4) $$y^2 = bx \quad \text{and} \quad x^2 = ay.$$

Let D be their point of intersection. Then the perpendiculars $x = DH$ and $y = DT$ satisfy (4) and hence (3).

Next, Omar considers the equation (2), in which N is a given number. He constructs a rectangular block with base e^2 and height Ne. Now he has to construct a cube equal to this block. In the case $N = 2$ this is just the well-known Greek problem of "doubling the cube". Hippokrates of Chios had proved that this problem can be reduced to the problem of finding two mean proportionals x and y between two given line segments a and b. Omar Khayyam proceeds just so. He solves the auxiliary problem (3) with $a = e$ and $b = Ne$, and he proves that the first intermediate x is the side of the required cube.

All this is well-known from Greek texts. According to Eutokios, the solution of (3) by means of the intersection of two parabolae is due to Menaichmos.

Next, Omar considers six types of cubic equations in which a binomial is equated to a monomial, namely

(5) $$x^3 + ax = b$$

(6) $$x^3 + b = ax$$

(7) $$x^3 = ax + b$$

(8) $$x^3 + ax^2 = b$$

(9) $$x^3 + b = ax^2$$

(10) $$x^3 = ax^2 + bx.$$

In Omar's terminology, the equation (5) is written as
"A cube and (a number of) sides are equal to a number".

Omar first constructs a square c^2 equal to the given number b, and next a block with base c^2 and height h equal to the given number b. This means, as he has explained earlier, that the block with sides c, c, and h is made equal to a block with sides e, e, and be, where e is the unity of length and b be the given number on the right hand side of equation (5). Thus, the equation (5) can be written in the homogeneous form

(11) $$x^3 + c^2 x = c^2 h$$

in which $c = AB$ and $h = BC$ are given line segments.

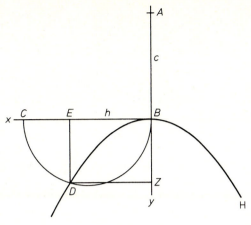

Fig. 9

To solve this equation geometrically, Omar constructs a parabola (see Fig. 9) having its summit at B, its axis being BZ and its "parameter" $AB=c$. Next he describes a semi-circle on the diameter $BC=h$. The semi-circle necessarily has a point of intersection D with the parabola. From D one draws perpendiculars DZ and DE to BZ and BC. Omar now proves that $DZ=x$ solves the equation (11).

In modern terminology, let $x=DZ$ and $y=DE$ be the coordinates of D. The equation of the parabola is

$$(12) \qquad\qquad x^2 = yc,$$

or, in Omar's own words: "The square of DZ will be equal to the product of BZ and AB". The equation of the circle is

$$(13) \qquad\qquad y^2 = x(h-x)$$

which Omar writes as a proportion
   "BE is to ED as ED is to EC".
   Just so, (12) is written as a proportion:
   "AB is to BE as BE is to ED".
   From these two proportions Omar concludes that $EB=x$ is a solution, and the only solution of his problem.
   Just so, Omar writes the equation (6) in the homogeneous form

$$(14) \qquad\qquad x^3 + c^2 h = c^2 x$$

and he solves it by intersecting the parabola

$$(15) \qquad\qquad yc = x^2$$

with the hyperbola

(16) $$y^2 = x(x - h).$$

The third equation (7) is solved in the same way, the only difference being the sign of the constant term b.

The next equation (8) is written as

(17) $$x^3 + ax^2 = s^3$$

where a and s are known line segments. Omar solves it by intersecting the hyperbola

$$xy = s^2$$

with the parabola

$$s(x + a) = y^2.$$

This solution is unnecessarily complicated, because it requires a preliminary solution of the equation

$$s^3 = b$$

by means of two parabolae. It would be much simpler to set $b = cd^2$ and to intersect the parabola

$$x^2 = cy$$

with the hyperbola

$$(x + a)y = d^2.$$

The next type (9) is solved by a similar method. Once more, the constant term b is made equal to a cube s^3. Omar notes that in this case the solution is not always possible.

The last type (10) is reduced to

(18) $$x^3 = ax^2 + ac^2$$

and solved by intersecting the hyperbola

$$xy = ac$$

with the parabola

$$y^2 = a(x - a).$$

Next, Omar discusses seven types of quadrinomial equations, namely

(19) $$x^3 + ax^2 + bx = c$$

(20) $$x^3 + ax^2 + c = bx$$

(21) $$x^3 + bx + c = ax^2$$

(22) $$c + bx + ax^2 = x^3$$

(23) $$x^3 + ax^2 = bx + c$$

(24) $$x^3 + bx = ax^2 + c$$

(25) $$x^3 + c = ax^2 + bx.$$

The methods of solution are the same as in the trinomial cases. To solve (19) one uses a circle and a hyperbola, to solve (20) two hyperbolae, and so on.

After this, Omar discusses equations in which terms like $1/x$, $1/x^2$, and $1/x^3$ occur. His first example is

$$x^3 = 10/x^3.$$

Multiplying both sides by x^3, one obtains

$$(x^3)^2 = 10$$

and hence

$$x^3 = \sqrt{10},$$

or in Omar's own words, as translated by Woepcke:

"Donc la racine de dix sera le cube cherché".

Omar notes that the equation

$$x^2 = a/x^3$$

cannot be solved by the methods exposed by him, because it requires the insertion of four mean proportionals between two given lines, as Ibn al-Haitham has proved.

Omar Khayyam was not the first to solve cubic equations by means of intersections of conics. At the end of his treatise he says that someone has told him that Muhammad ibn al-Lait Abu al-Jud was the author of a treatise in which he reduced the solution of cubic equations to conic sections, without however treating all cases. In particular, he taught the solution of type (21) by the intersection of a parabola and a hyperbola. On pages 84–85 of Woepcke's translation of the algebra of Omar Khayyam, the solution of (21) by Abu al-Jud is described.

Omar Khayyam on Ratios

On page 251 of his book "Geschichte der Mathematik im Mittelalter", A.P. Juschkewitsch has drawn the attention to two very remarkable passages on ratios in Omar Khayyam's commentary "Discussion of Difficulties of Euclid" (edited by Erani, Teheran 1936). Mrs. Yvonne Dold had the kindness to translate the two passages directly from Arabic into English. In what follows, I shall reproduce her translation.

As Juschkewitsch notes, Omar Khayyam states that Euclid's definition of proportion is correct, but that it is not a true definition of the notion ratio. The true meaning of a ratio is found in the process of measuring one magnitude by another magnitude. I shall now explain what this means.

Omar Khayyam defines a proportion of four magnitudes

(26) $$A:B=C:D$$

as follows:

> All multiples of the first are cut off from the second until a rest remains less than the first, and likewise all multiples of the third are cut off from the fourth until a rest remains less than the third. And the number of multiples of the first on the second is like the number of multiples of the third on the fourth. Moreover we cut off all multiples of the rest of the second from the first until a rest remains less than the rest of the second, and likewise all multiples of the rest of the fourth are cut off from the third until a rest remains less than the rest of the fourth. And the number of multiples of the rest of the second is like the number of multiples of the rest of the fourth. Likewise we cut off from the rest of the second all multiples of the rest of the first and we cut off from the rest of the fourth all multiples of the rest of the third. And the number of both is equal. Likewise we cut off all multiples of the rests one from the other according to the first part as we explained. And the number of every rest from the first and the second is like the number of its corresponding from the third and the fourth *ad infinitum*. Thus the ratio of the first to the second is inevitably as the ratio of the third to the fourth. And this is the true proportionality in the geometrical manner.

The process described here is what the Greeks call Antanairesis or Anthyphairesis: the continued mutual subtraction of two quantities A and B from each other. The smaller of the two, say B, is subtracted from A as often as possible, leaving a remainder R_1 less than B:

$$R_1 = A - q_1 B.$$

Next, R_1 is subtracted from B as often as possible:

$$R_2 = B - q_2 R_1$$

and so on. The integer quotients

$$q_1, q_2, \ldots$$

define the ratio $A:B$ in the following sense: If C and D have the same quotients as A and B, the proportion (26) holds.

From a passage on the Topica of Aristotle we know that this definition of the equality of ratios was used by the Greeks before Euclid. See O. Becker: Eudoxos-Studien I, Quellen und Studien Gesch. der Math. B 2, p. 311–333 (1933), or D.H. Fowler. Ratio in Early Greek Mathematics, Bulletin American Math. Soc. (New Series) 1, p. 807–848 (1979).

Omar Khayyam also defines the relation

$$A:B>C:D$$

by comparing the sequences of quotients

$$q_1, q_2, \ldots \quad \text{and} \quad q'_1, q'_2, \ldots$$

defining the two ratios. If m is the first index for which q_m differs from q'_m and if

$$q_m < q'_m \quad \text{for odd } m$$

or

$$q_m > q'_m \quad \text{for even } m,$$

then $A:B$ is larger than $C:D$.

In a series of theorems Omar Khayyam proves that his definition of the equality of ratios is equivalent to that of Euclid.

Next, al-Khayyam defines the multiplication of ratios, thus filling a logical gap in Euclid's Elements. See A.P. Juschkewitsch: Geschichte der Math. im Mittelalter, p. 253–254.

In his definition of the multiplication of ratios, al-Khayyam assumes the existence of a fourth proportional D to given quantities A, B, C. He justifies this assumption as follows: There always exist quantities M and N such that

$$C:M > A:B$$

and

$$C:N < A:B.$$

Now, because of the unlimited divisibility of continuous quantities, there must be a D between M and N such that

$$C:D = A:B.$$

Oscar Becker has noted that the same kind of argument, namely: "Where larger and smaller (quantities) exist, equals also exist" was also used by Greek commentators of Aristotle. See O. Becker: Eudoxos-Studien II. Warum haben die Griechen die Existenz der vierten Proportionale angenommen? Quellen und Studien Gesch. Math. B 2, p. 369–387.

Omar Khayyam also raises the question whether ratios can be regarded as a kind of "number" in a larger sense. He writes:

Then there is the question about the ratio of the magnitudes: is it inherent the number according to her nature, or a logical consequence of the number, or is it connected with the number by something that follows from its nature without the need of any external factor?

Omar Khayyam leaves this "philosophical" question unanswered, but later Arabic authors such as Nasir ad-Din at-Tusi consider all ratios as "numbers". See page 255 of the book of Juschkewitsch.

Chapter 2
Algebra in Italy

This chapter will be divided into three parts:
A. From Leonardo da Pisa to Luca Pacioli
B. Master Dardi of Pisa
C. The Solution of Cubic and Biquadratic Equations

Part A
From Leonardo da Pisa to Luca Pacioli

The methods of *al-jabr* and *al-muqabala* were made known in Italy first by the Latin translation of the algebra of al-Khwarizmi by *Gerard of Cremona*, and next by the work of *Leonardo da Pisa* (called Fibonacci). Leonardo was followed by several other writers of arithmetical textbooks, of which *Luca Pacioli* is best known. Before discussing the work of these authors, I shall first explain how the need for such textbooks was created by the economic conditions of the Italian merchants. In my exposition I shall gladly make use of the contents of a very interesting lecture entitled "The Contributions of the Italian Renaissance to European Mathematics", presented by Warren Van Egmond at a symposium held at Cortona in April 1983.

The Connection Between Trade and Civilization in Medieval Italy

In the early Middle Ages trade was mainly based on barter rather than on the exchange of money. Long-distance commerce was in the hands of travelling merchants, who exchanged their cargoes of goods at local fairs and markets. Major centres of this trade were Venice, Genova, and Pisa. Merchants set out by sea for the Arabic ports of North Africa and the Near East, carrying timber, wool, and other products from the West and bringing back in exchange fine silks, spices, jewels, and other precious goods.

In the thirteenth century the character of this economy changed radically. Improvements in navigation eased the dangers of sea travel. The increased circulation of coins made European economy predominantly monetary. The invention of letters of credit, bills of exchange, accounting and bookkeeping made possible the rise of banking and international finance. All of these developments worked together to create a new class of merchants, who lived in the major manufacturing and trading centres. They bought goods and shipped them to other representatives of the same company in other cities, for instance in the East, where the goods were sold or traded for other goods. Control over this vast network of representatives was maintained through a constant exchange of letters, bills, and reports. Im most cases the central office was based in one of the cities of Italy, such as Lucca, Siena, or Florence.

The life of a sedentary merchant was far different from that of his travelling predecessor. The early medieval merchant was a small tradesman, carrying his inventory in his head or on a scrap of paper. He calculated on his fingers or on a small abacus. On the other hand, the sedentary merchants and bankers wrote and received letters, bills of exchange, reports, orders, and so on. They had to calculate prices, to compute payments, to figure profits and losses.

For all these operations, they needed an efficient system of writing numbers and performing written calculations. The Roman numbers were too cumbersome: the Hindu-Arabic number system was much more efficient. The credit for developing this number system and adapting it to merchant practices belongs to a particular group of men, the so called "abbacists".

According to Warren Van Egmond, whose exposition I am following here, one has to distinguish between the Latin word *abacus*, which denotes a calculating board, and the Italian word *abbaco*, which usually means "practical arithmetic".

Life and Work of Fibonacci

The life of Leonardo da Pisa is well known from the introduction of his most famous work "Liber abbaci" (1202). In what follows, I shall follow the excellent description of his life and work in Kurt Vogel's article FIBONACCI in the Dictionary of Scientific Biography.

Leonardo was a member of the Bonacci family, hence he calls himself "filio Bonacci", which was shortened to Fibonacci. His father, a secretary of the republic of Pisa, was entrusted around 1192 with the direction of the Pisan trading company in Bugia (now Bougie), Algeria. He expected his son Leonardo to become a merchant, therefore he brought him to Algeria. Here Leonardo learned how to calculate with Hindu-Arabic numerals. His business trips took him to Egypt, Syria, Byzantium, Sicily, and southern France.

Around 1200 Leonardo returned to Pisa. During the next twenty-five years he composed several works. Five of these are preserved:

1. Liber abbaci (1202, revised 1228),
2. Practica geometriae (1220),
3. a book entitled "Flos" (1225),
4. a letter to the philosopher Theodorus, who lived in Sicily at the court of the Hohenstaufen emperor Frederick II,
5. Liber quadratorum (1225).

A treatise on Book X of Euclid's "Elements", containing a numerical treatment of the irrationalities which Euclid had demonstrated by lines and areas, is unfortunately lost.

Leonardo's importance was recognized at the court of Frederick II. In Leonardo's writings several names of scholars living at this court in Sicily are mentioned, including the astrologer Michael Scotus, whom Dante banished to hell (Inferno XX, 115), the philosopher Theodorus, and the mathematician John of Palermo. About 1225, when Frederick II held court at Pisa, the astronomer Dominicus presented Leonardo to the emperor. On that occasion, John of Palermo proposed several problems, which Leonardo solved promptly.

The first problem was, to find a number x such that $x^2 + 5$ and $x^2 - 5$ are square numbers. A solution, namely

$$x = 3\tfrac{5}{12}, \quad x^2 + 5 = (4\tfrac{1}{12})^2, \quad x^2 - 5 = (2\tfrac{7}{12})^2$$

was presented without proof in the book "Flos", which Leonardo sent to Frederick II. In the "Liber quadratorum" the solution was deduced by a method, which will be explained in the course of the present chapter.

The second problem proposed to Leonardo was the solution of the cubic equation

(1) $x^3 + 2x^2 + 10x = 20.$

In the book "Flos", Leonardo proved that the solution is neither an integer, nor a fraction, nor one of the irrationalities defined in Book X of the Elements of Euclid. He presented an approximate solution in sexagesimal form as

$$1; 22, \ 7, 42, 33, \ 4, 40.$$

According to Vogel (p. 610 of the article FIBONACCI) the 40 is too large by about $1\tfrac{1}{2}$, so Leonardo's accuracy is admirable. If he had applied the method of "double false position" explained by himself in the "Liber abbaci", that is, the method of linear interpolation between a smaller value x_1 and a larger value x_2, he would have obtained a too small approximation. It is possible that he used the so-called Horner method. This method, adapted to the sexagesimal system, consists in putting $x = 1 + y_1$ and obtaining an equation for y_1, next putting $60y_1 = 22 + y_2$ and obtaining an equation for y_2, and so on.

The history of the Horner method is very complicated. In principle, the method was known already to the author of the Chinese treatise "Nine Chapters of the Mathematical Art" (*Chiu Chang Suan Shu*), who lived in the Han-period, i.e. between -150 and $+150$. See J. Needham: Science and Civilization in China (Cambridge 1959), p. 126–127. The method was also known to the Arabic mathematician Jamshid al-Kashi. See P. Luckey: Die Rechenkunst des Ğamšīd b. Mas'ūd al-Kašī (Wiesbaden 1951). The method was later rediscovered by Paolo Ruffini (1804) and W.G. Horner (1819). See F. Cajori: A History of Mathematics, p. 271.

In 1240, the republic of Pisa awarded the "serious and learned Master Leonardo Bigolli" a yearly salary of 20 pounds silver "in addition to the usual allowances, in recognition of his usefulness to the city and its citizens through his teaching and devoted services".

We shall now discuss the extant works of Leonardo.

1. The "Liber Abbaci"

The Italian masters of computation were called "maestri d'abbaco". In this sense the title of Leonardo's most influential work is to be understood. It appeared first in 1202. To the second edition of 1228 "new material has been added, and superfluous removed". It was edited by Baldassare Boncompagni in Vol. 1 of the "Scritti di Leonardo Pisano" (Roma 1857). A summary of the 15 chapters of the "Liber abbaci" was given by Kurt Vogel in his article FIBO-NACCI in the Dictionary of Scientific Biography.

In Chapters 1–7 the Hindu-Arabic numerals are introduced, and methods of calculation with integers and fractions are taught.

Chapters 8–11 contain problems of concern to merchants. A remarkable playful problem is the "problem of the 30 birds". A man buys 30 birds: patridges, doves, and sparrows. A patridge costs 3 silver coins, a dove 2, and a sparrow $\frac{1}{2}$. He pays with 30 coins. How many patridges, doves, and sparrows does he buy?

The problem is, to solve the pair of equations

$$x + y + z = 30$$
$$3x + 2y + \tfrac{1}{2}z = 30$$

in positive integers x, y, z. The only solution is $x = 3$, $y = 5$, $z = 22$.

This problem is a variant of the "problem of 100 birds", which is found in Chinese, Indian, and Arabic sources. See Joh. Tropfke: Geschichte der Elementarmathematik I, fourth edition (by Kurt Vogel and others), p. 613–616 (1980).

Chapters 12 and 13 contain several types of recreational problems, some leading to a linear equation, others to two or three linear equations with two or three unknowns. For instance, we find on pages 228–243 a sequence of problems concerning "buying a horse". Leonardo begins with a simple case of two persons. One says to the other: "If you give me one-third of your cash, I can buy the horse." The other replies: "If you give me a quarter of your cash, I can buy the horse." If s is the price of the horse, we have two linear equations with two unknowns x and y:

$$x + 1/3\, y = s$$
$$y + 1/4\, x = s.$$

The problem is indeterminate, since s is not given. The solution in smallest integers is given as

$$x = (3 - 1) \times 4 \quad = 8$$
$$y = (4 - 1) \times 3 \quad = 9$$
$$s = 3 \times 4 - 1 \times 1 = 11.$$

Another case leads to 3 equations in 3 unknowns:

$$x+\tfrac{1}{3}(y+z)=s$$

(2)
$$y+\tfrac{1}{4}(x+z)=s$$

$$z+\tfrac{1}{5}(x+y)=s.$$

To solve these equations, Leonardo introduces a new unknown

(3)
$$x+y+z=t.$$

Subtracting each of the three equations from (3), one obtains

$$\tfrac{2}{3}(y+z)=\tfrac{3}{4}(x+z)=\tfrac{4}{5}(x+y)=t-s=D$$

hence

$$y+z=3/2\,D$$

$$x+z=4/3\,D$$

$$x+y=5/4\,D.$$

In order to obtain an integer solution, Leonardo puts $D=24$, thus obtaining

$$y+z=36$$

$$x+z=32$$

$$x+y=30$$

$$x=13,\quad y=17,\quad z=19.$$

This solution of the equations (2) had already been obtained by Diophantos: Arithmetica I 24.

The same problem of "buying a horse" occurs in a book of al-Karaji and in other Arabic and Byzantine sources. See J. Tropfke: Geschichte der Elementarmathematik I (4th edition, by K. Vogel and others), p. 608–609.

An original invention of Leonardo is the "series of Fibonacci"

$$0, 1, 1, 2, 3, 5, 8, 13, 21, \ldots$$

in which each term is the sum of the two preceding terms. Leonardo obtained it as the solution of the problem: How many pairs of rabbits can be produced from a single pair in a year if each pair begets a new pair every month, which from the second month on becomes productive, and if death does not occur?

Chapter 14 is devoted to calculations with square roots and cube roots. Leonardo begins by presenting some theorems from Euclid's Book II in numerical form, omitting the proofs, "because they are all in Euclid". For square roots he has the well-known approximation

$$\sqrt{a^2+r} \sim a+\frac{r}{2a}.$$

For the cube root Leonardo presents a first approximation

(4) $$\sqrt[3]{a^3+r} \sim a+\frac{r}{(a+1)^3-a^3}=a_1$$

and next a second approximation

(5) $$a_2=a_1+\frac{r_1}{3a_1(a+1)} \qquad \text{with } r_1=a-a_1^3.$$

According to Vogel (*DSB*, Article FIBONACCI, p. 608) the first approximation (4) was already known to al-Nasawi. In fact, it is a simple application of the rule of "double false position". As for the second approximation, Leonardo says: "I have invented this mode of finding roots."

Examples for his operations with radicals are

(6) $$\sqrt{4+\sqrt{7}}+\sqrt{4-\sqrt{7}}=\sqrt{14}$$

and

(7) $$4+\sqrt[4]{10}=\sqrt{16+\sqrt{10}+8\sqrt[4]{10}}.$$

Chapter 15 is very interesting. In a first section, Leonardo solves the pair of equations

(8) $$6:x=y:9$$

(9) $$x+y=21$$

as follows. From (8) he finds

$$xy=54$$

and next, using Euclid II, 5

$$\left(\frac{x-y}{2}\right)^2=\left(\frac{x+y}{2}\right)^2-xy=\left(\frac{21}{2}\right)^2-54=\frac{225}{4},$$

hence $x-y=15$, $x=18$, $y=3$.

In a second section, Leonardo presents applications of the Theorem of Pythagoras. For instance, he solves the problem: On the line joining the basis of two towers of given heights and given distance there is a spring which shall be equally distant from the tops of the towers. Leonardo gives a numerical as well as a geometrical solution (Liber abbaci, p. 398).

The third, most extensive section (p. 406–459) contains a systematic treatment of linear and quadratic equations. Citing "Maumeth", i.e. Muhammad ben Mūsā al-Khwārizmī, Leonardo solves the six normal forms

$$ax^2 = bx$$
$$ax^2 = c$$
$$bx = c$$
$$ax^2 + bx = c$$
$$ax^2 + c = bx \quad \text{(two solutions)}$$
$$ax^2 = bx + c.$$

The unknown quantity x is called *radix*, its square *quadratus* or *census*, and the constant term c *numerus*. The methods of solution are illustrated by numerous examples.

The first example of a mixed quadratic equation is just the same as in the algebra of al-Khwārizmī, namely

"census et decem radicis equantur 39"

or

$$x^2 + 10x = 39.$$

The solution is illustrated by a drawing (see Fig. 10):

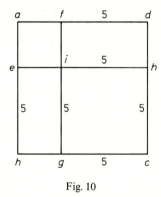

Fig. 10

In other examples, one has to divide 10 into two parts x and $10 - x$ satisfying an auxiliary condition such as

$$\frac{x}{10 - x} + \frac{10 - x}{x} = \sqrt{5}.$$

Leonardo also includes equations that can be reduced to quadratic equations. Thus, the set of equations

(10)
$$y = \frac{10}{x}$$

(11)
$$z = \frac{y^2}{x}$$

(12)
$$z^2 = x^2 + y^2$$

leads to a quadratic equation for x^4:

(13)
$$x^8 + 100x^4 = 10000.$$

2. The "Practica geometriae"

This work by Leonardo is extant in nine manuscripts, of which two are in Paris and four in Rome. In his edition, Boncompagni used only one of the vatican manuscripts. From Vogel's article FIBONACCI I quote:

In his work Leonardo does not wish to present only measurement problems for the layman; in addition, for those with scientific interests, he solves geometrical problems according to the method of proof. Therefore, the models are, on the one hand, Hero and the Agrimensores, and Euclid and Archimedes on the other. Leonardo had studied the *Liber embadorum* of Plato of Tivoli (1145) especially closely and took from it large sections and individual problems with the same numerical values. This work of Plato was a translation of the geometry of Savasorda (Abraham bar Hiyya), written in Hebrew, which in turn reproduced Arabic knowledge of the subject.

The *Practica* is divided into eight chapters (distinctiones), which are preceded by an introduction. In the latter the basic concepts are explained, as are the postulates and axioms of Euclid and the linear and surface measures current in Pisa.

In the first chapter, the proposition of Book II of the Elements of Euclid are recalled.

In the second chapter

the duplication of the cube by Archytas, Philo of Byzantium, and Plato, which are reported by Eutocius, are demonstrated, without reference to their source. The solutions of Plato and Archytas, Leonardo took from the *Verba filiorum* of the Banū Mūsā, a work translated by Gerard of Cremona. That of Philo appears also in Jordanus de Nemore's *De triangulis*, and probably both Leonardo and Jordanus took it from a common source. See M. Clagett, Archimedes in the Middle Ages I, p. 224 and 658–660.

The third chapter provides a treatment (with exact demonstrations) of the calculation of segments and surfaces of plane figures: the triangle, the square, the rectangle, rhomboids (*rumboides*), trapezoids (*figurae quae habent capita abscisa*), polygons, and the circle; for the circle, applying the Archimedean polygon of ninety-six sides, π is determined as $864:275 \sim 3.141818\ldots$.

For the surveyor who does not understand the Ptolemaic procedure of determining half-chords from given arcs, appropriate instructions and a table of chords are provided. This is the only place where the term *sinus versus arcus*, certainly borrowed from Arabic trigonometry, appears. The fourth chapter is devoted to the division of surfaces; it is a reworking of the *Liber embadorum*, which ultimately derives from Euclid's lost *Book on Divisions of Figures*; the latter can be reconstructed (see Archibald) from the texts of Plato of Tivoli and of Leonardo and from that of an Arabic version.

In the sixth chapter Leonardo discusses volumes, including those of regular polyhedrons, in connection with which he refers to the propositions of book XIV of Euclid.

The seventh chapter contains the calculation of the heights of tall objects, for example, of a tree, and gives the rules of surveying based on the similarity of triangles; in these cases angles are obtained by means of a quadrant.

The eighth chapter presents what Leonardo termed "geometrical subtleties" (*subtilitates*) in the preface to the *Liber abbaci*. Among those included is the calculation of the sides of the pentagon and the decagon from the diameter of the circumscribed and inscribed circles... .

3. The Book "Flos"

In the section "Life and Work of Fibonacci" we have already discussed Leonardo's solutions of two problems proposed to him by Giovanni da Palermo, namely: to find a number x such that x^2+5 and x^2-5 are squares, and to solve the cubic equation (1). Leonardo published his solutions in a book entitled "Flos", which he sent to Frederick II. We shall discuss his solution presently.

In addition to these solutions, the book contains some examples of indeterminate problems. Most of these had also been treated in the "Liber Abbaci". In some cases, negative solutions were interpreted as debts.

4. The Letter to Theodorus

The principal subject of this letter (*Scritti di Leonardo Pisano* II, p. 247–252) is the "Problem of the 100 birds", a variant of which had been discussed already in the *Liber abbaci*. In the letter, Leonardo develops a general method for the solution of indeterminate problems.

A geometrical problem follows. A regular pentagon is inscribed in a equilateral triangle. The solution is carried through to the point where a quadratic equation is reached, and then a sexagesimal approximation is presented.

5. The "Liber quadratorum"

The main subject of this book is the solution of the pair of Diophantine equations

$$x^2+5=y^2$$
$$x^2-5=z^2.$$

As we have seen, this was one of the problems which Giovanni da Palermo had proposed to Leonardo. Leonardo first generalizes the problem to

$$x^2+C=y^2,$$
$$x^2-C=z^2.$$

(14)

If x^2 and C form a solution of this problem, Leonardo calls the number C *congruum* and the square x^2 *quadratus congruentus*.

The pair of equations (14) is solved as follows. Adding, one obtains

(15) $$2x^2 = y^2 + z^2.$$

By the substitution $y = u + v$ and $z = u - v$ this equation can be reduced to

$$x^2 = u^2 + v^2.$$

This is the equation of Pythagorean triples. Its solution is well known:

$$x = a^2 + b^2, \quad u = 2ab, \quad v = b^2 - a^2.$$

If a and b are both odd, this solution can be divided by 2:

$$x = \frac{a^2 + b^2}{2}, \quad u = ab, \quad v = \frac{b^2 - a^2}{2}.$$

Thus, Leonardo obtains the theorem:
If a and b are relatively prime integers, and $b > a$, one has
(i) If a and b are both odd, then $C = ab(b-a)(b+a)$ is a congruum, and its congruent square is

$$x^2 = \left(\frac{a^2 + b^2}{2}\right)^2.$$

(ii) If a is odd and b even or conversely, then $C = 4ab(b-a)(b+a)$ is a congruum, and its congruent square is

$$x^2 = (a^2 + b^2)^2.$$

For $a = 1$ and $b = 9$, Leonardo finds

$$C = 720 = 5 \times 12^2$$

and

$$x = 41, \quad y = 49, \quad z = 31.$$

Dividing x, y, z by 12, Leonardo obtains a solution of his problem with $C = 5$, namely

$$x = 3 + \tfrac{5}{12}, \quad y = 4 + \tfrac{1}{12}, \quad z = 2 + \tfrac{7}{12}.$$

By the same method, Leonardo obtained solutions for other values of C. His successors calculated solutions for still more values of C. See Raffaella Franci: Numeri congruo-congruenti in codici dei secoli XIV e XV, Bollettino di Storia delle Scienze Matematiche 4, p. 3–23 (1984).

Leonardo's method differs from that of Abu Ǧa'far Muhammad ibn al-Husain, who also solved the same problem. Abu Ǧa'far's treatise "On the Construction of Rectangular Triangles with Rational Sides", in which his solution is contained, was translated into French by F. Woepcke in his paper "Recherches sur plusieurs ouvrages de Leonardo da Pisa III: Traduction d'un traité par Alhoçain", Atti dell' Accademia Pontificia dei Nuovi Lincei 14, p. 301–324 and 345–356 (1861).

For a general evaluation of Leonardo's abilities and his sources see pages 611–612 of Vogel's article FIBONACCI. Vogel is quite right: "With Leonardo, a new epoch in Western mathematics began."

Three Florentine Abbacists

In a very interesting paper entitled "Maestro Benedetto e la Storia dell'Algebra", Historia Mathematica 10, p. 297–317 (1983), Raffaela Franci and Laura Toti Rigatelli have discussed the work of *Maestro Benedetto* and two of his predecessors living in Florence in the fourteenth century, namely *Maestro Biaggio* and *Antonio Mazzinghi*. I shall now summarize their work.

1. Maestro Benedetto

In 1463, Benedetto of Florence completed his great work "Trattato di praticha d'arismetica", consisting of 500 large pergament pages. For us, the most interesting parts of this work are the books 13, 14, and 15, which deal with algebraic equations.

Benedetto starts with the well-known "reghola de algebra amuchable", that is, with the solution of the six types of linear and quadratic equations

$$x^2 = px \quad x^2 + px = q$$
$$x^2 = q \qquad x^2 + q = px$$
$$px = q \qquad\quad x^2 = px + q.$$

According to Franci and Toti Rigatelli (Historia Math. 10, p. 300) this part of Benedetto's text is a literal Italian translation of a Latin translation of the algebra of al-Khwarizmi.

Next, Benedetto introduces the well-known names for the powers of x, with suitable abbreviations such as

$$x^2 = \text{censo} \qquad\qquad = c$$
$$x^3 = \text{cubo} \qquad\qquad = b$$
$$x^4 = \text{censo di censo} = cc$$

and he presents rules of multiplication for these powers and their inverses and for radicals like $\sqrt{a}$ and $\sqrt[3]{a}$.

Benedetto next presents a long list of equations which can either be reduced to quadratic equations or directly solved by radicals, for instance

10. $x^3 + px^2 = qx$

which can be reduced to

$$x^2 + px = q,$$

or

15. $x^4 = px$

which can be solved as

$$x = \sqrt[3]{p}.$$

In book 15, he adds to this list three more types

37. $x^4 + px^2 = q$

38. $x^4 + q = px^2$

39. $x^4 = px^2 + q$

which can be solved first for x^2 and next for x.

2. Maestro Biaggio

In Book 14 of his Trattato, Benedetto presents a sequence of 140 numerical problems derived from a lost "Trattato di Praticha" written by the Florentine master Biaggio, who died circa 1340. Twenty-eight of these are mercantile problems. The others are theoretical: they lead to algebraic equations all belonging to the types solved by Benedetto in Book 13. One of these problems leads to an equation

$$\tfrac{1}{12}x^2 + (2 + \tfrac{1}{12})x + 12 = x$$

which, according to Biaggio followed by Benedetto, "non può essere". In fact, this equation has no real roots.

Another problem of Biaggio leads to an equation

$$x^4 + x^2 = 110.$$

The only positive solution is

$$x = \sqrt{10}.$$

3. Antonio Mazzinghi

In Book 15 of his Trattato, Maestro Benedetto has included a short bio-graphy of Maestro Antonio, who was a member of the family of the Mazzinghi. He had his atelier near Santa Trinità at Florence, and he became very famous, not only in arithmetic and geometry, but also in astrology and music. He died circa 1390.

From Antonio's treatise "Fioretti" Benedetto quotes several rather difficult problems, such as

To find numbers in continuous proportions such that their sum is 19 and the sum of their squares 40.

In modern notation, the conditions for the three numbers x, y, z are

$$x + y + z = 10$$

$$x : y = y : z$$

$$x^2 + y^2 + z^2 = 40.$$

From these equations one derives first

$$2xy + 2xz + 2yz = 100 - 40 = 60$$

$$xz = y^2$$

and next, replacing the term xz by y^2,

$$2(x + y + z)\, y = 60$$

$$y = 3$$

and finally

$$x + z = 7$$

$$x^2 + z^2 = 31$$

hence

$$x = \tfrac{1}{2}(7 + \sqrt{13}), \qquad z = \tfrac{1}{2}(7 - \sqrt{13}).$$

It seems that Antonio Mazzinghi was the first to introduce, besides the traditional name "cosa" for an unknown quantity, a special name for another unknown. One of his problems reads: to find two numbers such that their sum is 18 and the sum of their squares 27. He now assumes the first number to be "una cosa meno la radice d'alchuna quantità", and the second "una chosa più la radice d'alchuna quantità". That is, he supposes the two numbers to have the form

$$x - \sqrt{y} \quad \text{and} \quad x + \sqrt{y}.$$

I feel we cannot but admire the mathematical ability of Maestro Antonio. In their paper in Historia Math. 10, Franci and Toti Rigatelli conclude that many algebraic methods usually ascribed to Luca Pacioli were already used by the Florentine abbacists Biaggio, Antonio Mazzinghi, and Benedetto.

Two Anonymous Manuscripts

Two Italian manuscripts from the Biblioteca Nazionale di Firenze, namely

$$\text{Fond. Princ. II.V. 152}$$

and
$$\text{Conv. Sopp. G.7. 1135,}$$

contain very interesting methods for solving cubic equations. By the kindness of Raffaella Franci I have seen a preprint of her paper "Contributi alla risoluzione dell'equazione di 3° grado nel XIV secolo", to be published in Festschrift Gericke (Steiner-Verlag, Wiesbaden).

The first manuscript just mentioned contains a sequence of 22 equations which can be reduced to quadratic or pure cubic equations. The sequence ends with three types we have met already in the work of Benedetto, namely

20. $$ax^4 = bx^2 + c$$

21. $$ax^4 + c = bx^2$$

22. $$ax^4 + bx^2 = c.$$

Next comes an extremely interesting passage concerning cubic equations of the three types

23. $$ax^3 + bx^2 = c$$

24. $$ax^3 = bx^2 + c$$

25. $$ax^3 + c = bx^2.$$

For the solution of these equations, the author presents prescriptions as follows. First the equations are divided by a and thus reduced to the case $a = 1$. In the case 23. we now have to solve an equation of the form

$$x^3 + px^2 = q.$$

Next, $x + \frac{1}{3}p$ is introduced as a new unknown, which I shall call y. Thus one obtains an equation of the form

(16) $$y^3 = ry + s.$$

If y is a solution of the equation (16), y is called "the cube root of s with supplement r" (la radice cubica di s con l'aggiunta di r). For instance, the cube root of 44 with supplement 5 is 4, because

$$4^3 = 5 \times 4 + 44.$$

If the supplement is left indetermined, it is always possible to find the root: one has just to find a number y such that y^3 exceeds s. But if the supplement is given, one has to proceed by trial and error.

It may happen that in (16) the term s is negative. For instance, the equation

$$2x^3 + 36x^2 = 704$$

leads to $x + 6 = y$ and

(17) $y^3 = 108y - 80.$

In this case, the author says that one has to find the cube root of "debito 80" with supplement 108. Thus, *negative numbers are regarded as debts*. The solution of (17), found by trial and error, is $y = 10$.

The second manuscript discusses the same twenty-two types of problems reducible to linear and quadratic equations as the first, and next two more types, corresponding to types 23 and 25 of the first manuscript, but illustrated by different numerical examples, namely by two examples for type 23:

$$24x^3 + 81x^2 = \;516 \quad \text{(solution } x = 2\text{)}$$

$$3x^3 + 27x^2 = 1620 \quad \text{(solution } x = 6\text{)}$$

and by one example for type 25:

$$16x^2 = x^3 + 576 \quad \text{(solution } x = 12\text{)}.$$

The method of solution is the same as in the first manuscript.

We now come to the best known author of this period:

Luca Pacioli

Luca Pacioli's main work "Summa de arithmetica, geometria, proportioni e proportionalità", written in Italian in 1487, was printed at Venice in 1494. It was very influential.

As compared with Fibonacci, Luca has a simpler algebraic notation. He denotes the square root by R or $R2$, the cube root by $R3$, the fourth power root by $R4$ or RR (Radix Radix). The unknown in an equation is denoted by co. (cosa), its square by ce. (censo), its cube by cu. (cubo), its fourth power by ce.ce. (censo censo). If a second unknown is introduced, it is called "quantità". For addition and subtraction the signs p and m are used. Thus,

$$R\,V\,40\,\tilde{m}\,R\,320$$

means $\sqrt{40 - \sqrt{320}}$. The letter V indicates that the root has to be extracted from the whole expression that follows ($V = U =$ Universale).

At the end of his book, Luca Pacioli states that for equations, in which

	numero, cosa e cubo	$(n, x$ and $x^3)$
or	numero, censo e cubo	$(n, x^2$ and $x^3)$
or	numero, cubo e censo de censo	$(n, x^3$ and $x^4)$

occur, "it has not been possible until now to form general rules".

Soon afterwards, these cubic and biquadratic equations were solved by Scipione del Ferro, Tartaglia, Cardano, and Ferrari, as we shall see in Part C of the present chapter. Before discussing the work of these sixteenth century algebrists, we have first to consider a very remarkable abbacist living at Pisa in the fourteenth century.

Part B
Master Dardi of Pisa

A little known work of Master Dardi of Pisa entitled "Aliabraa argibra" has been examined by Warren Van Egmond in a recent paper "The Algebra of Master Dardi of Pisa" in Historia Mathematica 10, p. 399–421 (1983). I think I can do no better than quote his Summary:

This article presents a summary list of 198 different types of equations and their rules of solution found in an algebra text of the 14th century, which is attributed to an otherwise unknown master Dardi of Pisa. The text is especially noteworthy for its unusual length, its adept handling of complex equations involving radicals and powers up to the 12th degree, and its correct solution of four irreducible cubic and quartic equations.

The importance of Dardi's treatise for the history of algebra has already be pointed out by Guillaume Libri in the second volume of his "Histoire des Sciences Mathématiques en Italie" in a footnote on page 519.

Dardi's treatise is extant in three Italian copies and one Hebrew translation, which was written in 1473 by Mordechai Finzi at Mantua. Finzi states that Dardi wrote his treatise in 1344.

Dardi's list of problems begins with the six well-known types of linear and quadratic equations. Next come cubic and biquadratic equations such as

7. $$a x^3 = n$$

8. $$a x^3 = b x$$

12. $$a x^4 = b x$$

which can be solved by extracting cube roots or square roots. After this, Dardi presents a long sequence of equations (most of them involving radicals) which can be reduced to quadratic or pure cubic equations, for instance

37. $$n = a x + \sqrt{b x}.$$

Four special cases inserted between nos. 182 and 183 deserve special attention, because they involve irreducible mixed cubic and biquadratic equations. In modern notation, these four equations can be written as

(1) $$cx + bx^2 + ax^3 \qquad = n$$

(2) $$dx + cx^2 + bx^3 + ax^4 = n$$

(3) $$dx + cx^2 + ax^4 = n + bx^3$$

(4) $$dx + ax^4 = n + cx^2 + bx^3.$$

Dardi presents rules for the solution of these equations. However, as he himself admits, his rules are valid only in the special cases considered, not in general. In all cases, he first instructs us to divide all coefficients by a, so that, for instance, Equation (1) is reduced to the simpler form

(1') $$x^3 + bx^2 + cx = n.$$

The solution of (1') is given as

(5) $$x = \sqrt[3]{(c/b)^3 + n} - c/b.$$

Now the question arises: How did Dardi arrive at his rule (5)?

I don't know the answer, but I may venture a hypothesis. In the work of al-Khwārizmī and also in that of Leonardo da Pisa a quadratic equation

$$x^2 + bx = c$$

is solved by adding to both sides a constant such that the left hand side becomes a complete square $(x + b/2)^2$. Now let us try to add to both sides of (1') a constant such that the left hand side becomes a cube $(x + L)^3$. Under what conditions does this procedure work?

Al-Khwārizmī and Leonardo da Pisa both illustrate the formula for the square of $(x + b/2)$ by a drawing like our Fig. 10. Now let us try to draw a

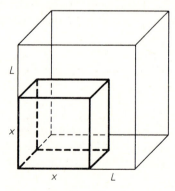

Fig. 11

similar diagram for $(x+L)^3$ (see Fig. 11). Geometrically, it is clear that the larger cube can be divided into eight parts:

one part x^3, called "il cubo",

three blocks $x^2 L$,

three blocks $x L^2$,

and one cube L^3.

If one now supposes that these parts are just equal to the three terms on the left hand side of (1') plus a constant term to be added to both sides, one obtains three conditions that have to be satisfied, namely

(A)   Added Term $= L^3$

(B)   "Number of Cose" c $= 3 L^2$

(C)   "Number of Squares" $b = 3 L$.

Condition (A) can always be satisfied by a suitable choice of the added term. Dividing (B) by (C), one obtains

$$L = c/b.$$

Now Dardi's example is chosen in such a way that $L = c/b$ satisfies (B) and (C). The Equation (1') can now be written as

$$(x+L)^3 = n + L^3$$

and its solution is

$$x = \sqrt[3]{n+L^3} - L = \sqrt[3]{n+(c/b)^3} - c/b,$$

in full accordance with Dardi's solution.

Dardi's example of an equation (1) is

(1') $x^3 + 60 x^2 + 1200 x = 4000.$

In this case it is completely clear that the equation can be written as

$$(x+20)^3 = 4000 + 8000 = 12000$$

and solved by extracting a cube root.

Dardi's example (1') comes from a loan problem. The same problem is also found in a manuscript entitled "Trattato d'Abaco" by Piero della Francesca, the famous painter. See Gino Arrighi: „Note di Algebra di Piero della Francesca", Physics 9, p. 421–424 (1967). In this Trattato we find three loan problems, two of which also occur, with exactly the same numerical data, in Dardi's treatise.

The first problem reads:

Someone lends to another one 100 Lira, and after 3 years receives 150 Lira with annual capitalization of the interest. One asks at what monthly rate of interest the loan was given.

The monthly rate of interest is expressed in denarii pro Lira, 1 Lira being 20×12 denarii. Thus, if the monthly rate of interest is x denarii pro 1 Lira, the annual interest is $12 x$ denarii pro Lira, and the rate of interest is $x/20$. Thus one obtains the equation

$$100 \left(1 + \frac{x}{20}\right)^3 = 150$$

or, if one multiplies by 80

$$(x + 20)^3 = 12000$$

which is just Dardi's equation (1′).

The second problem is similar. If the creditor gets 160 Lira after 4 years, the equation for x becomes

$$(x + 20)^4 = 256000$$

or

$$x^4 + 80 x^3 + 2400 x^2 + 32000 x = 96000$$

which is Dardi's equation (2′). It can be solved by extracting a fourth power root.

The originator of these three problems seems to be Dardi.

Dardi's examples of Equations (3) and (4) are not of the same type. His examples read

(3′) $x^4 + 28 x^2 + 720 x = 20 x^3 + 1800$

and

(4′) $x^4 + 1120 x = 20 x^3 + 12 x^2 + 2800.$

The problems leading to Equations (3′) and (4′) are both of the same form, namely

Problem P. To divide 10 *into two parts such that their product divided by their difference is* $\sqrt{g}$,

with $g = 18$ in (3′) and $g = 28$ in (4′).

If one part is called x and the other $10 - x$, we have

(6) $$\frac{x(10 - x)}{x - (10 - x)} = \sqrt{g}$$

with $g = 18$ or $g = 28$. In the first case we have

$$x^2(10 - x)^2 = 18(2x - 10)^2$$

or

(7) $$x^4 - 20 x^3 + 28 x^2 + 720 x = 1800,$$

which is Dardi's equation (3′).

In the second case, we have

$$x^2(10-x)^2 = 28(2x-10)^2$$

or

(8) $$x^4 - 20x^3 - 12x^2 + 1120x = 2800$$

which is Dardi's equation (4′).

Dardi's solution of Equation (3) with $a=1$ reads

(9) $$x = \sqrt[4]{(c/4)^2 + n} + b/4 - \sqrt{d/2b}$$

and his solution of (4) with $a=1$ reads just so.

How did Dardi arrive at this curious formula? Once more, I may venture a hypothesis.

The Equation (6) may be written as a quadratic equation

$$(2x-10)\sqrt{g} = 10x - x^2$$

or

(10) $$x^2 - 2(5-\sqrt{g})x = 10\sqrt{g}.$$

The solution of (10) is

$$x = 5 - \sqrt{g} + \sqrt{(5-\sqrt{g})^2 + 10\sqrt{g}}$$

or

(11) $$x = 5 - \sqrt{g} + \sqrt{25 + g}.$$

On the other hand, the squaring of (6) yields

$$x^2(10-x)^2 = g(2x-10)^2$$

or

(12) $$x^4 + (100 - 4g)x^2 + 40gx = 20x^3 + 100g$$

and if we write this as

$$x^4 + cx^2 + dx = bx^3 + n$$

we have

(13) $$b = 20, \quad c = 100 - 4g, \quad d = 40g, \quad n = 100g.$$

Now, if one inserts the values (13) in Dardi's formula (9), one obtains

$$x = \sqrt[4]{(25-g)^2 + 100g} + 5 - \sqrt{g},$$

which accords with (11).

Dardi's problem was: How can I write the solution (11) in a form like (9), in which not special numbers like 5 and g (equal to 18 or 28) occur, but only expressions which can be calculated from the coefficients (13)?

Now let us try to find out how Dardi solved his problem. Let's consider the three terms of (11) separately.

The first term 5 was obtained by halving the 10 given in the Problem P, and $b=20$ was found by doubling this term. So, the first term in (11) can only be generalized to $b/4$. Thus, the second term in (9) is explained.

The second term $\sqrt{g}$ in (11) is the square root of the given number g (equal to 18 or 28), and the coefficient d is $40\,g$, so if we divide d by $2b=40$ we obtain just g. Hence the third term in (9) is

$$-\sqrt{d/2\,b} = -\sqrt{g}.$$

The first term in (9) is more difficult to explain. I suppose that Dardi wanted to obtain an expression analogous to his formula (5). In (5) the first term is a third root of "something plus n", where n is the constant term on the right hand side of his equation (1). Just so, Dardi had obtained a solution of (2), in which a fourth root of "something plus n" occurred. So I suppose that Dardi wanted to write the third term of (11) in the form

$$\sqrt[4]{\text{something plus } n}.$$

He reached his aim by making the "something" equal to $(c/4)^2$, for we have

$$\sqrt[4]{(c/4)^2 + n} = \sqrt[4]{(25-g)^2 + 100\,g}$$
$$= \sqrt[4]{(25+g)^2} = \sqrt{25+g}.$$

Of course, may hypothesis is not proved, but it does at least explain the facts. In any case we cannot but admire Dardi's skill in finding his formula (9) without the help of our algebraic notation.

Part C
The Solution of Cubic and Biquadratic Equations

The solution of the general cubic equation is due to the sixteenth century Italian algebrists Scipione del Ferro, Tartaglia, and Cardano. I shall now describe their work, which is of fundamental importance for the history of algebra.

Scipione del Ferro

The general cubic equation

$$x^3 + a\,x^2 + b\,x + c = 0$$

can be reduced, by introducing a new variable

$$x' = x + \tfrac{1}{3}a,$$

to the simpler form

$$x^3 + px + q = 0.$$

If only positive coefficients and positive values of x are admitted, there are 3 types

(1) $$x^3 + px = q$$

(2) $$x^3 = px + q$$

(3) $$x^3 + q = px.$$

The first to solve equation (1) was Scipione del Ferro, who was professor at the university of Bologna until his death in 1526. According to E. Bortolotti (Periodico di Matematica, serie 4, Vol. 5, 1925, p. 147–184) he actually solved all three problems (1), (2), (3), but this is not quite certain.

The fundamental idea underlying the solution of (3) is very simple. I shall follow Cardano's explanation of the method, given in his book "Ars Magna, sive de regulis algebraicis" (first printed in Nürnberg 1545), Chapter 11. Cardano starts with the example

(4) $$x^3 + 6x = 20.$$

Cardano expresses this equation in the language of his "rhetorical algebra", as

"Let a cube and six time its side equal 20".

Cardano's idea is, to solve the equation (4) by putting

(5) $$x = u - v.$$

Cardano expresses this in geometric terminology as follows. He represents our u by a line segment AC, and our v by CK, and then he says: "Marking off BC equal to CK, I say that, if this is done, the remaining line AB is equal to GH" (that is, to our x).

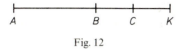

Fig. 12

Substituting $x = u - v$ into (4), one obtains

$$x^3 + 6x = (u - v)^3 + 6(u - v)$$
$$= (u^3 - v^3) - 3uv(u - v) + 6(u - v) = 20.$$

Now u and v are subjected to the following conditions

(6) $$u^3 - v^3 = 20$$

(7) $$3uv = 6.$$

Then it follows that $x = u - v$ satisfies the required equation

$$x^3 + 6x = 20.$$

In Cardano's geometrical terminology the reduction of $(u-v)^3$ to

$$(u^3 - v^3) - 3uv(u-v)$$

is very cumbersome, but the fundamental idea is the same.

It is easy to determine u and v from the conditions (6) and (7). From (7) one finds

$$uv = 2$$

hence

$$u^3 v^3 = 8.$$

Now the difference and the product of the two cubes u^3 and v^3 are known, and one finds

$$u^3 = \sqrt{108} + 10$$
$$v^3 = \sqrt{108} - 10,$$

so u and v are cube roots of known numbers, and we have

$$x = \sqrt[3]{\sqrt{108} + 10} - \sqrt[3]{\sqrt{108} - 10}.$$

Cardano formulates this as a general rule:

Cube one third the "number of sides" (i.e. one-third the coefficient of x). Add to it the square of one-half the constant of the equation, and take the square root of the whole. You will put this twice, and to one of the two you add one-half the number you have squared and from the other you subtract one half the same. You will then have a binomium ($\sqrt{108} + 10$) and its apotome ($\sqrt{108} - 10$). Then, subtracting the cube root of the apotome from the cube root of the binomium, the remainder is the required side.

Scipione del Ferro never published his solution: he only told a few friends. Among these was his pupil Antonio Maria Fiore, from Venice. With his entrance into the scene, a dramatic development begins.

Tartaglia and Cardano

At this time challenge disputes, often for considerable sums of money, were a normal form of competition in the learned world. A mathematics teacher in Venice, *Niccolò Tartaglia*, born in Brescia in 1499, was very successful in these contests and won several prizes. Tartaglia, "the stutterer", was his nickname; his real name seems to have been Niccolò Fontana.

In 1535, Tartaglia was challenged to a problem-solving contest by Scipione del Ferro's pupil Fiore. There were to be 30 questions, and the loser was to pay for 30 banquets. Tartaglia prepared a variety of problems, but Fiore had only one arrow to his bow: all his problems were equations of the form (1). The night between February 12 and 13; shortly before the experiation of the allotted time, Tartaglia had an inspiration. He discovered the method of solution of equation (1), and solved all 30 problems within a few hours. In the contest, Fiore proved unable to solve most problems of Tartaglia and was declared the loser. The honour alone was satisfaction enough to Tartaglia, and he renounced the 30 banquets.

Early in 1539, another actor entered the scene: *Gerolamo Cardano*, a famous medical doctor, astrologer, philosopher, and mathematician, who lived in Milan. His life has been vividly described by Oystein Ore in his book "Cardano, The Gambling Scholar" (Princeton 1953, reprinted by Dover, New York 1965). Cardano had heard of Tartaglia's discovery, and he approached Tartaglia, sending the bookseller Zuan Antonio da Bassano to Venice as an intermediary. As Tartaglia would not reveal his method, Cardano urged him to come to Milan and to stay in his house; he promised Tartaglia to introduce him to the marchese Alfonso d'Avalos, the military commander of Milan.

Tartaglia accepted the invitation. He had made some military inventions, and he was eager to show them to the marchese.

After Tartaglia's arrival at Milan, Cardano persuaded him to reveal the secret of the solution of "the cube and the cose", that is, of the equation (1). Cardano swore an oath that he would never publish Tartaglia's discovery. The oath was sworn, according to Tartaglia's account, on March 25, 1539.

Right after Tartaglia's visit, Cardano succeeded in extending the method of solution of Equation (1) to the other types (2) and (3). In these two cases, one has to write the solution as $u+v$ instead of $u-v$. For the rest, the calculation is just the same. Thus in case (2) one has

$$x^3 - px = q$$
$$x = u+v$$
$$x^3 - px = u^3 + v^3 + 3uv(u+v) - p(u+v) = q$$
$$3uv = p$$
$$u^3 + v^3 = q$$
$$u^3 = \tfrac{1}{2}q + w$$
$$v^3 = \tfrac{1}{2}q - w$$

with

(8)
$$w = \sqrt{(\tfrac{1}{2}q)^2 - (\tfrac{1}{3}p)^3}$$

and hence

(9)
$$x = u + v = \sqrt[3]{\tfrac{1}{2}q + w} + \sqrt[3]{\tfrac{1}{2}q - w}.$$

But now a new difficulty arises. The difference under the root sign in (8) may become negative. In this case, the so-called "casus irreducibilis", no real square root is possible. Yet, the equation can be solved in real numbers. In the "casus irreducibilis" there are even three real roots. They may be obtained as follows.

In the "casus irreducibilis" we may write $w = \sqrt{-c}$. The two expressions $\frac{1}{2}q + w$ and $\frac{1}{2}q - w$ are complex conjugates. For the first cube root in (9) we have three possibilities: the root may be multiplied by a cube root of unity. Because of the condition $3uv = p$, the second cube root in (9) must be chosen complex conjugate to the first, so the sum is always real, no matter which of the three possibilities is chosen.

Cardano knew about this difficulty, for in Chapter 1 of his "Ars Magna" he presents a complete discussion of the number of positive or negative roots of cubic equations of the types (1), (2), (3). He knows that in the cases (2) and (3), if $(\frac{1}{3}p)^3$ exceeds $(\frac{1}{2}q)^2$, there are all in all three real roots. However, in explaining the solution by means of cube roots, he carefully avoids the "casus irreducibilis". In all of his examples, w is always a root of a positive number, and there is only one (positive) root.

In the "casus irreducibilis" one has to extract a square root from a negative number. Such square roots, which we now call "imaginary", occur also in another chapter of the "Ars Magna". In Chapter 37, Cardano poses the problem: To divide 10 in two parts, the product of which is 40. He writes:

> It is clear that this case is impossible. Nevertheless, we will work thus: We divide 10 into two equal parts, making each 5. These we square, making 25. Subtract 40, if you will, from the 25 thus produced, as I showed you in the chapter on operations in the sixth book, leaving a remainder of -15, the square root of which added to or subtracted from 5 gives parts the product of which is 40. These will be $5 + \sqrt{-15}$ and $5 - \sqrt{-15}$.

Cardano next verifies that the two numbers thus obtained satisfy the required conditions. He writes:

> Putting aside the mental tortures involved, multiply $5 + \sqrt{-15}$ by $5 - \sqrt{-15}$, making $25 - (-15)$, which is $+15$. Hence this product is 40. ... This is truly sophisticated. ... (Translated by T.R. Witmer: The Great Art or The Rules of Algebra by Girolamo Cardano, M.I.T. Press, Cambridge, Mass. 1968.)

As far as I know, Cardano was the first to introduce complex numbers $a + \sqrt{-b}$ into algebra, but he had serious misgivings about it.

Lodovico Ferrari

In 1536, a youth of 14 years came into Cardano's household as a servant. He learned mathematics and developed into an eminent mathematician, Cardano's friend and secretary.

Ferrari discovered that the general equation of degree 4 can be reduced to a cubic equation and hence be solved by means of square roots and cube roots. Cardano explained Ferrari's method in Chapter 39 of his "Ars Magna", and he stated that "it is Lodovico Ferrari's, who gave it to me on my request".

Cardano's exposition of the method starts with a theorem about squares and rectangles, which he explains as follows:

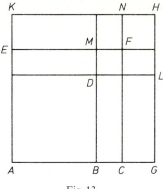

Fig. 13

"Let the square AF be divided into two squares AD and DF, and two supplements, DC and DE, and let me add the gnomon KFG around it in order to complete the whole square AH (see Fig. 13). I say that this gnomon will consist of GC^2 plus twice the added line $GC \times CA$, for FG is $GC \times CF$, from the definition given at the beginning of the second book of the Elements, and CF equals CA by the definition of a square. Since, according to I, 43 of the Elements, KF equals FG, the two surfaces GF and FK consist of $GC \times 2CA$, and GC^2 equals FH, according to the corollary to II, 4 of the Elements. Hence the proposition is clear. If, therefore, AD equals x^4, and CD and DE [each] equal $3x^2$, and DF equals 9, BA will equal x^2 and BC will necessarily equal 3. Since, therefore, we shall wish to add more squares to DC and DE, these will be CL and KM. In order to complete the whole square LMN is necessary. This, as has been demonstrated, consists of the square of GC [plus $2GC \times BC$], one-half the [original] number of squares, for CL is the surface produced by $GC \times AB$, as has been shown, and AB is x^2 because we assumed that AD is x^4 and, therefore, FL and MN are made up of $GC \times CB$, according to I, 42 of the Elements. Hence the surface LMN (this is the number to be added) is $GC \times 2BC$ (that is, times the coefficient of x^2, which is 6) plus GC times itself (that is, times the added number of squares). This demonstration is our own."

Let me explain this, using our modern algebraic notation. If we put $AB = s$, $BC = a$, and $CG = b$, the theorem proved by Cardano is equivalent to the identity

$$(s + a + b)^2 = (s + a)^2 + 2sb + 2ab + b^2.$$

In the appliation of this identity to the solution of the biquadratic equation, Cardano takes for $s = AB$ the square of the unknown x, so that he obtains

(10) $$(x^2 + a + b)^2 = (x^2 + a)^2 + 2x^2 b + 2ab + b^2.$$

In a biquadratic equation, the term x^3 can always be made to disappear, so only terms with x^4, x^2, x, and a constant term remain. As an example, Cardano considers the equation

(11) $$x^4 + 6x^2 + 36 = 60x.$$

In order to reduce the left-hand side to a square $(x^2 + a)^2$, he adds $6x^2$ to both sides, thus obtaining

(12) $$(x^2 + 6)^2 = 6x^2 + 60x.$$

Next he says:

Now if $6x^2 + 60x$ had a square root, we would have the solution. But it does not. Hence there must be added to both sides alike enough squares and a number so that on one side there is a trinomium with a root and on the other the same.

This means: if $6x^2 + 60x$ would be the square of a binomium $px + q$, we would extract square roots from both sides of (12). But since $6x^2 + 60x$ is not a complete square, we have to add a term $2bx^2$ and a constant term to both sides in order to obtain complete squares on both sides. Putting $a = 6$ in the identity (10), Cardano has an identity

$$(x^2 + 6 + b)^2 = (x^2 + 6)^2 + 2bx^2 + 12b + b^2.$$

So, if one adds

$$2bx^2 + 12b + b^2$$

to both sides of (12), one obtains

(13) $$(x^2 + 6 + b)^2 = (6x^2 + 60x) + (2bx^2 + 12b + b^2)$$
$$= (2b + 6)x^2 + 60x + (b^2 + 12b).$$

Now b is chosen in such a way that the right hand side of (13) becomes a complete square of a binomium $px + q$. The condition for this is

$$(2b + 6)(b^2 + 12b) = 30^2$$

or

$$2b^3 + 30b^2 + 72b = 900$$

or

(14) $$b^3 + 15b^2 + 36b = 450.$$

This is a cubic equation for b, which can be solved by the method explained in an earlier chapter of Cardano's book. The result is

$$b = \sqrt[3]{190 + \sqrt{33903}} + \sqrt[3]{190 - \sqrt{33903}} - 5.$$

Now the right-hand side of (13) is a complete square, and one can extract square roots from both sides, thus obtaining a quadratic equation for x.

Cardano and Ferrari now were in a awkward position. They had made extremely important discoveries, but they could not publish them, because

Cardano had sworn an oath by the Sacred Gospel never to publish Tartaglia's solution of the cubic equation, which formed the basis of their common work.

In the year 1543 Cardano and Ferrari decided to go to Bologna and ask Annibale della Nave whether there was any truth in the rumours that Scipione del Ferro had discovered the solution of the cubic equation even before Tartaglia. They were well received and readily given permission to examine the posthumous papers of Scipione, in which the solution was clearly explained.

Now Cardano decided to publish the solution of the cubic and biquadratic equation in his book "Ars Magna" (1545), stating clearly that the solution of equation (1) had been discovered by Scipione del Ferro and rediscovered by Tartaglia, that he himself had extended the solution to equations (2) and (3), and that the solution of the biquadratic equation was due to Ferrari.

Tartaglia was furious. The very next year he published the story of the oath, with all details, including the text of the oath.

Rafael Bombelli

Rafael Bombelli was the author of a very influential work in three books entitled l'Algebra. It was first printed in Venice 1572, and next in Bologna 1579.

Bombelli admired Cardano's "Ars Magna", but he felt that Cardano had not been clear in his exposition ("ma nel dire fù oscuro"). So he decided to write a treatise that would enable a beginner to master the subject without the aid of any other book.

Book 1 of Bombelli's Algebra deals with the calculation of radicals, in particular of square roots and cube roots. Very remarkable is his approximation of square roots by continued fractions. To approximate $\sqrt{2}$, Bombelli writes

(1)
$$\sqrt{2} = 1 + \frac{1}{y}.$$

From this he finds

(2)
$$y = 1 + \sqrt{2}.$$

By adding 1 to both sides of (1), one obtains

(3)
$$y = 2 + \frac{1}{y}.$$

Substituting (3) into (1), Bombelli finds

$$\sqrt{2} = 1 + \frac{1}{2 + \frac{1}{y}}.$$

I shall follow the usage to write this continued fraction as

$$1+\frac{1}{2+}\frac{1}{y}.$$

Continuing in this way, Bombelli obtains an infinite continued fraction

$$\sqrt{2}=1+\frac{1}{2+}\frac{1}{2+}\frac{1}{2+}\cdots.$$

If, after a finite number of steps, one neglects $1/y$, one obtains an approximation of $\sqrt{2}$, for instance

$$1+\tfrac{1}{2}=\tfrac{3}{2},$$

or

$$1+\frac{1}{2+}\frac{1}{2}=\frac{7}{5},$$

and so on.

Bombelli applies the same method to other square roots such as $\sqrt{13}$. He
obtains a first approximation

$$\sqrt{13}\sim3+\tfrac{4}{6}=3\tfrac{2}{3}$$

and a second approximation, replacing the 6 in the denominator by $6+\tfrac{4}{6}$,

$$\sqrt{13}\sim3+\frac{4}{6+}\frac{4}{6}=3\frac{3}{5}.$$

Chapter 2 of Bombelli's Algebra deals with the solution of equations up to
degree 4. For the cubic and biquadratic equations he follows Cardano. In
contrast to Cardano, he fully discusses the "casus irreducibilis". Solving the
equation

(4) $x^3=15x+4$

by the rule of Cardano, he finds

(5) $x=\sqrt[3]{2+\sqrt{-121}}+\sqrt[3]{2-\sqrt{-121}}.$

Following Cardano, Bombelli calls the imaginary roots "sophistic", but he
notes that the equation (4) is by no means impossible, for it has the root $x=4$.
He now investigates whether he can attach a meaning to the cube root of a
complex number. More precisely, he tries to equate the first cube root in (5) with
a complex number $p+\sqrt{-q}$:

(6) $\sqrt[3]{2+\sqrt{-121}}=p+\sqrt{-q}.$

This would yield

$$2+\sqrt{-121}=(p^3-3pq)+(3p^2-q)\sqrt{-q}.$$

This equation can be satisfied by putting

(7) $$2 = p^3 - 3pq$$

and

(8) $$\sqrt{-121} = (3p^2 - q)\sqrt{-q}.$$

Now if these two conditions are satisfied, we also have

(9) $$\sqrt[3]{2 - \sqrt{-121}} = p - \sqrt{-q}.$$

Multiplying (6) and (9), Bombelli obtains

$$\sqrt[3]{125} = p^2 + q$$

or

(10) $$q = 5 - p^2.$$

Substituting this into (7), one obtains a cubic equation for p:

(11) $$4p^3 - 15p = 2.$$

A solution of this equation is $p = 2$, and from (10) one has

$$q = 5 - 4 = 1,$$

so we have

$$\sqrt[3]{2 + \sqrt{-121}} = 2 + \sqrt{-1}$$

and

$$\sqrt[3]{2 - \sqrt{-121}} = 2 - \sqrt{-1},$$

hence

$$x = \sqrt[3]{2 + \sqrt{-121}} + \sqrt[3]{2 - \sqrt{-121}}$$
$$= (2 + \sqrt{-1}) + (2 - \sqrt{-1}) = 4.$$

After having found this result, Bombelli was very much satisfied. He writes: "At first, the thing seemed to me to be based more on sophism than on truth, but I searched until I found the proof."

Bombelli introduced a notation for what we call $+i$, namely *più di meno*, and for $-i$, *meno di meno*. He presented rules of calculation such as

meno di meno uia men di meno fà meno,

which means

$$(-i) \times (-i) = -1,$$

and he gives some examples of calculations involving complex numbers.

I shall give an example of Bombelli's notation. The expression

$$\sqrt[3]{2+\sqrt{-121}} = \sqrt[3]{2+11\,i},$$

which occurs in his solution of a cubic equation, is written as

R.c. $L\,2$ p. di m. 11$\lrcorner$.

Here R.c. means Radice cubica. The Letter L and the inverted L at the end play the role of brackets: the cube root is to be extracted from the whole expression between the L and the inverted L. The abbreviation p. di m. means *più di meno*.

Chapter 3
From Viète to Descartes

François Viète

For the life of François Viète (1540–1603) see the article VIÈTE by H.L.L. Busard in Dictionary of Scientific Biography XIV, p. 18–25.

Viète was born in Fontenay-le-Comte on the river Vendée. He studied law at the University of Poitiers, where he received a bachelor's degree in 1560. Four years later he entered the service of Antoinette d'Aubeterre as a secretary and educator of her daughter Cathérine de Parthenay. His lectures on the elements of geography and astronomy have been published in 1637 under the title "Principes de cosmographie". His "Harmonicon coeleste", a treatise in five books on Ptolemaic astronomy, is extant in several manuscripts.

In 1571, Viète began to publish his "Canon mathematicus, seu ad triangula cum appendicibus". The publication of the first two books, dealing with plane and spherical trigonometry, was finished in 1579. The last two books of the Canon, on astronomy, have not been published.

In 1573, Viète was appointed counselor to the parliament of Brittany at Rennes, and in 1580 he became "maître de requêtes" at Paris, an office attached to the parliament. In 1584 he was banished from the royal court, but in 1589 he was recalled by Henri III and became counselor of the parliament at Tours.

During the war against Spain, Viète served Henri IV by decoding intercepted letters written in a code. For details of his decoding see D. Kahn: The Code Breakers (New York 1968), p. 116–118.

Viète returned to Paris in 1594 and to Fontenay-le-Comte in 1597.

Most important for the history of algebra is Viète's "In artem analyticem Isagoge" (Tours, 1591). His aim was, to revive the method of analysis explained by Pappos in his great "Collection" and to combine it with the methods of Diophantos. To the two kinds of analysis mentioned by Pappos Viète added a third, which he called "rhetic" or "exegetic" (from ἐξηγέομαι = to lead, to show the way), and which he defined as the procedure by which an unknown magnitude is found by solving an equation.

Viète was the first to use letters not only for unknowns but also for known quantities. He used the consonants $B, C, D, \ldots$ to denote known quantities, and vowels $A, E, \ldots$ to denote unknowns.

In Chapter 3 of his "Isagoge", Viète explains his "Law of Homogeniety", according to which only magnitudes of "like genus" can be compared or added. Thus, where we would write a quadratic equation as

$$bx^2 + dx = z,$$

Viète writes

"*B* in *A* Quadratum, plùs *D* plano in *A*, aequari *Z* solido".

Namely, by Viète's Law of Homogeneity, if *A* (our *x*) and *B* (our *b*) are line segments, *D* must be a plane area and *Z* a volume. Hence he writes "*D* plano" and "*Z* solido".

This law implies a serious restriction of the algebraic formalism. As we have seen, Omar Khayyam managed to circumvent this restriction by introducing a unit of length *e*. We shall see that Descartes used the same trick.

In Chapter 4, Viète formulates the "canonical rules" of "species calculation", that is, of calculation with letters, as opposed to calculation with definite numbers.

In Chapter 5, Viète presents rules for solving equations. One operation called "antithesis" is the transfer of terms from one side of an equation to the other side, corresponding to what the Arabic algebrists call *al-jabr*. Another operation is the division of all terms of an equation by one and the same "species", and so on.

In 1593 Viète published his "*Zeteticorum libri quinque*" (five books on finding). In this work he explained the solution of several determinate and indeterminate problems. Some of the problems are taken from the "Arithmetica" of Diophantos. A typical example is the problem: to divide a number, which is a sum of two squares, into two other squares.

In two further treatises Viète discusses the geometrical solution of algebraic equations. In the first, entitled "*Effectionum geometricarum canonica recensio*", he shows that the solution of quadratic equations can be constructed using circles and straight lines only. For instance, in order to solve the equation

$$A^2 + AB = D^2 \quad \text{or} \quad A(A + B) = D^2,$$

Viète constructs two perpendicular line segments *B* and *D*, next he draws a semi-circle centered at the midpoint of *B*. The two remaining parts of the diameter are equal to *A* (see Fig. 14).

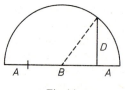

Fig. 14

In Viète's second treatise "Supplementum geometriae" (1593) he adds to Euclid's construction postulates for straight lines and circles one more postulate, namely: "To draw a straight line from a given point across any two lines (straight lines or a straight line and a circle) such that the intercept between the two lines is equal to a given distance". In the history of Greek mathematics, constructions based on this postulate are called "*neusis constructions*".

By means of this postulate, Viète first solves the problem of constructing two mean proportionals between two given line segments. The solution of this problem immediately yields the doubling of a cube. Next, Viète solves the trisection of an angle. By the same method he also constructs a regular heptagon inscribed in a circle. Finally he shows that all geometrical problems leading to cubic or biquadratic equations can be solved by means of neusis constructions.

In the same year 1593, Viète decided to publish book VIII of his "*Variorum de rebus mathematicis responsorum*". In Chapters 1–7 of this book he discusses, once more, the doubling of the cube, the trisection of an angle, and the construction of a regular heptagon. In Chapter 8 he considers the quadratix, in Chapter 11 the lunules that can be squared, and in Chapter 16 he presents a construction of the tangent at any point of an Archimedian spiral.

Most interesting is Chapter 18, in which π is represented as an infinite product. The area of a polygon of 4×2^n sides inscribed in a circle of radius 1 can be written as

$$\frac{2}{c_1 c_2 c_3 \ldots c_n}$$

with

$$c_1 = \sqrt{\tfrac{1}{2}}$$

$$c_2 = \sqrt{\tfrac{1}{2} + \tfrac{1}{2} c_1}$$

$$c_3 = \sqrt{\tfrac{1}{2} + \tfrac{1}{2} c_2}$$

and so on. Letting n go to infinity, one obtains

$$\pi = \frac{2}{c_1 c_2 c_3 \ldots}.$$

In 1593 the Dutch mathematician Adrianus Romanus proposed to all mathematicians the problem of solving a certain equation of degree 45. The ambassador of the Netherlands at the court of the French king Henri IV claimed that nobody in France would be able to solve this problem. The king thereupon informed Viète of the challenge. Viète saw that the equation was solved by the chord subtending an arc of 8 degrees in a circle of radius 1. Thus, the solution can be found by dividing the circumference into 45 equal parts. During the same audience, Viète presented one root of the equation, and the next day all 23 positive roots. He published his solution in 1595 in a treatise entitled "*Ad problema, quod omnibus mathematicis totius orbis construendum proposuit Adrianus Romanus, responsum*".

In 1615, after the death of Viète, his Scottish friend Alexander Anderson published in one volume two papers of Viète entitled "*De aequationem re-*

cognitione" and "*De equationem emendatione*". In the latter paper Viète discusses several methods of transforming equations. For instance, if one has one root D of an equation, one can obtain an equation of lower degree. For this, Viète presents several examples.

Another example. Let a cubic equation

(1) $$BA - A^3 = Z$$

be given, and let E satisfy the condition

(2) $$E^3 - Z = BE.$$

For us, this means that $-E$ is a root of the original equation (1), but Viète does not acknowledge negative roots. From (1) and (2) he concludes

$$A^3 + E^3 = B \cdot (A + E)$$

or

$$(A + E)(A^2 - AE + E^2) = B \cdot (A + E).$$

Now one can divide by $A + E$, and one obtains a quadratic equation for A.

In the same paper, Viète deals with the solution of biquadratic and cubic equations. He starts with a biquadratic equation

(3) $$A^4 = Z - BA.$$

If $A^2 E^2 + \frac{1}{4} E^4$ is added to both sides, one obtains

(4) $$(A^2 + \tfrac{1}{2} E^2)^2 = Z - BA + A^2 E^2 + \tfrac{1}{4} E^4.$$

The right hand side becomes a complete square, if E satisfies the equation

$$Z + \tfrac{1}{4} E^4 = B^2 / 4 E^2$$

or

(5) $$E^6 + 4 Z E^4 = B^2,$$

which is a cubic equation for E^2. Viète's method is essentially the same as that of Ferrari.

In Chapter 7 of the "Emendatione" a new method is taught for solving the cubic equation

(6) $$A^3 + 3BA = 2Z.$$

Viète introduces a new unknown E by the equation

(7) $$B = E(A + E).$$

Substituting (7) into (6) one obtains

$$A^3 + 3AE(A+E) = 2Z$$

and hence

(8)
$$(A+E)^3 = 2Z + E^3.$$

From (7) one can solve $A+E$ and substitute it into (8). One obtains a quadratic equation for E^3:

(9)
$$B^3 = 2ZE^3 + E^6$$

which can be solved for E^3 and hence for E:

(10)
$$E = \sqrt[3]{\sqrt{B^3 + Z^2} - Z}.$$

Now A can be formed from (7). In contrast to the method explained in Cardano's Ars Magna, one has to extract only one cube root. Yet, the final result is the same as in Cardano's method, for if one introduces another unknown $E' = A + E$, one has

$$B = E'(E' - A)$$

and one can derive, as before, a quadratic equation for E'^3:

$$B^3 = E'^6 - 2ZE'^3$$

from which one obtains

(11)
$$E' = \sqrt[3]{\sqrt{B^3 + Z^2} + Z}.$$

Now $A = E' - E$ is a differences of two cube roots, as in Cardano's Ars Magna.

Viète knows about the relation between the roots and the coefficients of an equation. In Chapter 10 of the "Emendatione" he formulates a theorem:

Si A cubus $\overline{-B-D-G}$ in A quad. $+B$ in $D+B$ in $G+D$ in G in A, aequatur B in D in G: A explicabilis est de quadlibet illarum trium B, D vel G.

This means: If

$$A^3 + (-B - D - G)A^2 + (BD + BG + DG)A = BDG$$

then A equals any one of the three quantities B, D, or G.

Another paper published by Anderson is entitled "Ad angulares sectiones theoremata $\kappa\alpha\theta o\lambda\iota\kappa\acute{\omega}\tau\epsilon\rho\alpha$" (Most General Theorems on Divisions of Angles). In this paper, Viète considers the trisection of an angle and uses it to obtain a trigonometrical solution of a cubic equation in the "Casus irreducibilis". If one puts, with an arbitrary radius R,

$$2R \cos \varphi = A$$
$$2R \cos 3\varphi = \pm B$$

one has the equation

(12) $$A^3 = 3R^2 A \pm R^2 B,$$

and every cubic equation having three real roots can be reduced to this form and solved by trigonometry.

Simon Stevin

Simon Stevin, born at Brugge in 1548, came to Leyden in 1582. Here he published several books on mathematics and mechanics, all in Dutch. In his opinion the Dutch language was, of all languages, the best to express ideas in general, and scientific ideas in particular. Several mathematical expressions, coined by Stevin, are still in use at Dutch schools. For instance, we still call mathematics "wiskunde", that is, the science of that what is certain (gewis).

Stevin was an excellent engineer. He built windmills, locks, and ports. Maurits, the prince of Orange, the great leader in the war against Spain, used Stevin as an adviser in buildung fortifications. Stevin's book on this subject, called "Stercktebouw", became very popular. It influenced the famous French builder of the fortifications Vauban.

In Stevin's "Wereldschrift" (World Script) he defended the Coperican system. His highly original books on mechanics were inspired by Archimedes.

Most influential was a booklet of 36 pages entitled "De Thiende" (The Tenth), first published in 1585. In the same year Stevin brought out a French translation: "La Disme". For a facsimile of the Dutch edition with English translation see "The Principal Works of Simon Stevin", Vol. 2 (Amsterdam 1958), p. 371–454.

In this booklet, Stevin denotes the units by ⓪, their tenth parts by ①, and so on. As an example, I shall reproduce his multiplication of 0.000378 by 0.54:

```
              ④   ⑤   ⑥
              3    7    8
                   5    4    ②
            ┌─────────────
       1    5    1    2
  1    8    9    0
  ┌────────────────────
  2    0    4    1    2
  ④   ⑤   ⑥   ⑦   ⑧
```

Decimal fractions were used by the Chinese and Arabs long before Stevin (see J. Tropfke: Geschichte der Elementarmathematik, 4th edition, p. 106 and 110–112), but it was Stevin who made them popular in the West.

Soon after Steving the decimal point came into use. For instance, on page 218 of Clavius' "Astrolabium tribus libris explicato" (Rome 1593) one finds the notation 46.5, and Napier, who published his table of logarithms in 1614, used the decimal point systematically.

Of course, the decimal notation can be used for all real numbers, whether rational or irrational. Like our engineers, Stevin does not make a distinction between rational and irrational numbers. He says right at the beginning of his book "L'arithmétique" (Leyden 1585):

> Nombre est cela, par lequel s'explique la quantité de chacune chose,

and

> Nombre n'est point quantité discontinue.... Il n'y a aucuns nombres irrationels, irréguliers, inexpliquable, ou sourds.

Thus, with one stroke, the classical restriction of "numbers" to integers (Euclid) or to rational fractions (Diophantos) was eliminated. For Stevin, the real numbers formed a continuum. His general notion of a real number was accepted, tacitly or explicitly, by all later scientists. According to Descartes, Leibniz, and Newton, every ratio of one line to another can be expressed by a "number". For the wording of their definitions see J. Tropfke: Geschichte der Elementarmathematik I, p. 137.

Stevin also accepted negative numbers, as did several of his predecessors. However, he did not accept imaginary solutions of equations, because "they don't help us in finding real solutions".

In his book "Stelreghel" (= Algebra) Stevin introduced several simplifications of the algebraic notation. Thus, he used $+$ and $-$ for addition and subtraction, M and D for multiplication and division, $\sqrt{\ }$ for square root, $\sqrt{③}$ for cube root, and so on.

Pierre de Fermat

For a survey of the life and work of Fermat see Michael S. Mahoney: The Mathematical Career of Pierre de Fermat (1601–1665), Princeton University Press 1973. A lively description of Fermat's brilliant work in number theory has been given by André Weil: Number Theory, An approach through history, Birkhäuser, Basel 1983. Here we shall be concerned only with Fermat's discovery of the method of Analytic Geometry.

Analytic geometry was invented, nearly simultanuously and independently, by Fermat and Descartes. The invention was not very difficult, but it was of fundamental importance for the development of geometry and algebra. Its primary aim was to solve geometrical problems by algebraic methods. Conversely, the method can also be used to apply geometrical methods to algebraic problems.

Fermat's exposition of the method of analytic geometry is explained in his "Introduction to Plane and Solid Loci". In January 1643, Fermat sent this treatise to his correspondent Pierre de Carcavi. It was published after the death of Fermat in his "Varia Opera" (1679), and again in a better edition in Vol. 1 of the "Oeuvres de Fermat" (1891).

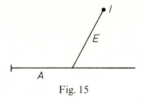

Fig. 15

What are "Plane and Solid Loci"? Fermat gives the following explanation:

Whenever the local endpoint of the unknown quantity describes a straight line or a circle, a plane locus results, and when it describes a parabola, hyperbola, or ellipse, a solid locus results (Mahoney's translation).

Let me explain this. In Fermat's "Introduction" a variable point I is determined by two (orthogonal or skew) coordinates, which he denotes by A and E, using Viète's notation (see Fig. 15). The "local endpoint" is the endpoint I of the ordinate E. So, according to Fermat's explanation, a "plane locus" is a circle or a straight line, and a "solid locus" is a conic section.

The first part of the "Introduction", dealing with plane loci, was finished in April 1636. This is clear from a letter to Mersenne, in which Fermat writes:

I have completely restored Apollonius' treatise *On Plane Loci*. Six years ago I gave it to Mr. Prades.... It is true that the prettiest and most difficult problem, which I had not yet solved, was missing. Now the treatise is complete in every point, and I can assure you that in all of geometry there is nothing comparable to these propositions (Mahoney's translation).

What was this "most difficult problem"? This is clear from a letter to Roberval, written in September 1636, in which Fermat first explains some applications of his method of coordinates, and next continues:

I have omitted the principal application of my method, which is for finding plane and solid loci. It has served me in particular for finding that plane locus that I earlier found so difficult: If from any number of given points straight lines are drawn to a (variable) point, and if the sum if the squares of the lines is equal to a given area, the point lies on a circumference given in position.

The solution of this "most difficult" problem is presented as Theorem II 5 of the "Introduction". Fermat first treats several special cases, in which the given points lie on a straight line. Next he deals with the case in which one of the given points (Q in Fig. 16) lies outside the line AE. To treat this case, he

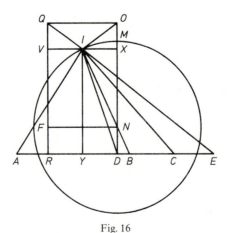

Fig. 16

introduces orthogonal coordinates for the point Q as well as for the variable point I. The coordinates of I are IX and IY. Now the problem is easy to solve, and there is no difficulty in extending the method to the general case.

In principle, the method of coordinates had already been used by Apollonios in his "Conica". See for this subject Otto Neugebauer: Apollonius-Studien, Quellen und Studien Gesch. der Math. B 2, p. 215–254. In the "Conica", a variable point on a conic section is determined by two line segments, usually called "abscissa" and "ordinate". Following modern usage, I shall denote the two line segments by x and y (see Fig. 17). If the point I varies on a conic section, there is a definite algebraic relation between x and y, which is called the "symptoma" of the curve.

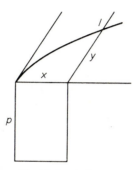

Fig. 17 Parabola

In the "Conica", the "symptoms" of the three conic sections are

$$\begin{array}{lll} \text{Parabola} & y^2 = px & \text{(see Fig. 17)} \\ \text{Hyperbola} & y^2 : x(a+x) = p : a \\ \text{Ellipse} & y^2 : x(a-x) = p : a. \end{array}$$

In Fermat's "Introduction", the equation of a straight line through the origin reads

$$D \cdot A = B \cdot E$$

(see Fig. 18). Fermat formulates this as a theorem:

If $D \cdot A = B \cdot E$, then the locus of the point I is a straight line.

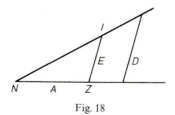

Fig. 18

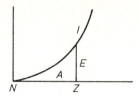

Fig. 19. Parabola

In the same way, the parabola is treated (see Fig. 19). Fermat shows:

If $Aq.$ (that is, the square of A) equals $D \cdot E$, the point I lies on a parabola.

The equation of the circle is written as

(13) $$Bq. - Aq. = Eq.$$

and Fermat proves that all equations containing Aq and Eq and A and E multiplied with given quantities may be reduced to this Equation (13), provided the angle NZI is right and the coefficient of $Aq.$ is equal to that of $Eq.$

At the end of his treatise, Fermat explains a general method to reduce any quadratic equation in x and y to one of the special forms

$$
\begin{aligned}
ax &= by &&\text{straight line} \\
xy &= b &&\text{hyperbola} \\
x^2 \pm xy &= ay^2 &&\text{pair of lines} \\
x^2 &= ay &&\text{parabola} \\
b^2 - x^2 &= y^2 &&\text{circle} \\
b^2 - x^2 &= ay^2 &&\text{ellipse} \\
b^2 + x^2 &= ay^2 &&\text{hyperbola.}
\end{aligned}
$$

Thus every quadratic equation in x and y represents a straight line or a conic section. The same result was proved, by a different method, by Descartes, as we shall see presently.

René Descartes

Our algebraic notation is mainly due to René Descartes (1596–1650). Descartes introduced this notation right at the beginning of his treatise "La géométrie", in which the principles of "analytic geometry" are explained. This treatise is a part of Descartes' great philosophical work "Discours de la Méthode" (1637). Descartes obviously considered his "Géométrie" as a standard example to elucidate his general considerations concerning the method of science.

The French text of "La Géométrie" has been published, together with an English translation, by David Eugene Smith and Marcia Latham in 1954. It was reprinted by Dover, New York.

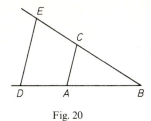

Fig. 20

Descartes begins by saying that "any problem in geometry can be reduced to such terms that a knowledge of the lengths of certain lines is sufficient for its construction". Now the standard operations in arithmetic are: addition, subtraction, multiplication, division, and the extraction of square roots. In geometry, after having chosen a fixed line called "unity", one can define five operations on line segments corresponding to the arithmetical operations. Of course, line segments can be added and subtracted. To multiply two line segments BD and BC, if AB is taken as unity, one has only to join the points A and C, and to draw DE parallel to CA; then BE is the required product (see Fig. 20). In other words: the product $ab = c$ is defined by the proportion

$$(14) \qquad\qquad\qquad e:a = b:c$$

where e is the unity.

One sees the advantage of this notation. In Greek geometry, the product of two line segments is an area: it cannot be added to a straight line segment. On the other hand, according to Descartes the product of two line segments is again a line segment. Just so, the quotient of two line segments is a line segment: if (14) holds, b is the quotient c/a.

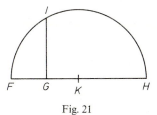

Fig. 21

The square root of a line segment is explained thus: One adds to GH a segment $FG = e$, one divides FH into two equal parts at K, one draws a semicircle FIH with centre K, and one erects a perpendicular GI to FH. The line segment GI is the required root (see Fig. 21).

Descartes now denotes his line segments by $a, b, \ldots,$ and he writes

$$a+b, a-b, ab, \frac{a}{b}, \sqrt{a}.$$

He also writes

$$\sqrt{aa+bb}$$

for the square root of a^2+b^2. Thus, all essentials of our algebraic notation have been established by Descartes.

In contrast to Stevin, Descartes does not introduce the notion "real number". For him, the quotient of two line segments is a line segment. Thus, he avoids all logical difficulties connected with irrational numbers.

Descartes next shows how the equations

$$z^2 = az + b^2$$
$$y^2 = -ay + b^2$$
$$z^2 = az - b^2$$

can be solved geometrically.

Next, Descartes turns to a kind of problems that had been discussed by Euclid and Apollonios, namely the "problem of three or four lines". If three straight lines are given in position, and if line segments are drawn from a variable point in given angles to those three lines, and if it is given that the rectangle on two of these line segments is in a given proportion to the square on the third; or if four straight lines are given, and if line segments are drawn from a variable point in given angles to the four lines, and if the rectangle on two of these line segments is in a given proportion to the rectangle on the two remaining lines, then it is required to prove that the point lies on a given conic section.

According to Pappos, Apollonios says in the third book of his treatise on the "locus of three or four lines", that Euclid had not solved this problem, and that he himself too had not been able to solve it completely, nor had anyone else. "This", says Descartes "led me to try to find out whether, by my own method, I could go as far as they had gone".

Descartes next states that in the case of three or four lines the required points lie all on one of the conic sections, or even in some cases on a circle or straight line. He now proceeds to prove this by his own method of coordinates.

Let AB, AD, EF, and GH be straight lines given in position (see Fig. 22). It is required to find points C such that the line segments CB, CD, CF, and CH drawn from C to the four given lines satisfy the condition that $CB \cdot CD$ be in a given proportion to $CF \cdot CH$:

(15) $$CB \cdot CD = \alpha \cdot CF \cdot CH.$$

Let this condition be satisfied, says Descartes, and let the segments AB and BC be called x and y. Thus, every point C is determined by two coordinates x and y, the angle ABC being given.

It is very strange that orthogonal coordinates are called, after the latinized name of Descartes, "Cartesian coordinates". Descartes himself does not suppose ABC to be a right angle.

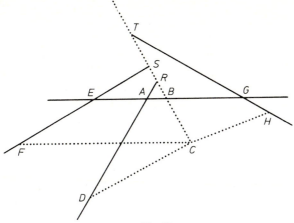

Fig. 22

Descartes next shows that all segments CB, CD, CF, and CH are *linear functions* of x and y. It follows that the Equation (15) is a quadratic equation for x and y. For every assumed value of y (or x) the corresponding value of x (or y) can be constructed by solving a quadratic equation. Thus, the required curve can be drawn, says Descartes.

In the second book of his Geometry, Descartes proceeds to investigate the nature of the curves thus obtained. His notations are rather clumsy. He reduces the Equation (15) to the form

(16)
$$y^2 = 2my - \frac{2n}{z}xy + \frac{bcfglx - bcfgxx}{ez^3 - cgzz}$$

and next, simplifying the notations a little, he solves for y:

(17)
$$y = m - \frac{n}{z}x + \sqrt{mm + ox + \frac{p}{m}xx}.$$

Introducing $y - m + \dfrac{n}{z}x = y'$ as a new coordinate, Descartes simplifies (17) to

(18)
$$y' = \sqrt{mm + ox + \frac{p}{m}xx}$$

and he shows, using some theorems from the first book of the Conica of Apollonios, that the curve is a conic section or a straight line.

Descartes' method can be applied to any curve determined by a quadratic equation. The final result is the same as that obtained by Fermat: Every quadratic equation in x and y determines a conic section, or in exceptional cases a straight line.

Chapter 4
The Predecessors of Galois

Modern algebra begins with Evariste Galois. With Galois, the character of algebra changed radically. Before Galois, the efforts of algebrists were mainly directed towards the solution of algebraic equations. Scipione dal Ferro, Tartaglia, and Cardano showed how to solve cubic equations, and Ferrari succeeded in solving equations of degree 4. Gauss proved that the cyclotomic equation

$$x^n - 1 = 0$$

can be completely solved by radicals, and that every algebraic equation can be solved by complex numbers $a + bi$. Galois, on the other hand, was the first to investigate the structure of fields and groups, and he showed that these two structures are closely connected. If one wants to know whether an equation can be solved by radicals, one has to analyse the structure of its Galois group. After Galois, the efforts of the leading algebrists were mainly directed towards the investigation of the structure of rings, fields, algebras, and the like.

The most important predecessors of Galois were Lagrange, Gauss, and Abel. The work of Gauss on algebraic equations will be discussed in Chapter 5. In the present chapter, we shall discuss the work of Waring, Vandermonde, Lagrange, Malfatti, Ruffini, Cauchy, and Abel on the solution of algebraic equations.

Waring

If an equation of degree n

$$x^n - a_1 x^{n-1} + a_2 x^{n-2} - \ldots + \ldots = 0$$

has n roots, it is well known since Viète that the coefficients of the equation are all equal to the elementary symmetric functions of the roots:

$$a_1 = x_1 + x_2 + \ldots + x_n$$

$$a_2 = x_1 x_2 + x_1 x_3 + \ldots + x_{n-1} x_n$$

etc.

In his treatise "Miscellanea analytica" (Cambridge 1762), Edward Waring has shown that all rational symmetric functions of the roots can be expressed as rational functions of the coefficients of the equation. He first derives expressions for the power sums

$$s_m = x_1^m + x_2^m + \ldots + x_n^m$$

and next for arbitrary symmetric polynomials.

In a later treatise "Meditationes algebraicae" (Oxford 1770) Waring derives another method for expressing symmetric polynomials. This method is the same we find in modern textbooks (see e.g. Weber's Lehrbuch der Algebra, second or third edition, p. 163–167).

Waring also investigates the solution of the "cyclotomic equation"

$$x^n - 1 = 0$$

and discusses the problem: to find equations that can be solved by sums of the form

$$x = \sqrt[m]{\alpha_1} + \sqrt[m]{\alpha_2} + \ldots + \sqrt[m]{\alpha_n}.$$

Thus, Waring is certainly one of the earliest predecessors of Galois theory.

Vandermonde

The mathematical work of Alexandre-Théophile Vandermonde has been discussed in a very interesting paper of Lebesgue: "L'oeuvre mathématique de Vandermonde", L'enseignement mathématique, New Series 1, p. 203–223 (1955).

In 1770, Vandermonde presented to the Paris Academy a memoir entitled "Sur la résolution des équations". Starting with the well known solution of quadratic and cubic equations, Vandermonde develops general principles upon which the solution of equations may be based. He writes the solution of the quadratic equation in the form

$$\tfrac{1}{2}[x_1 + x_2 + \sqrt{(x_1 - x_2)^2}].$$

Taking for the square root the two possible signs, one obtains the two roots. Next he rewrites his formula as

$$\tfrac{1}{2}[(x_1 + x_2) + \sqrt{(x_1 + x_2)^2 - 4x_1 x_2}],$$

thus introducing the elementary symmetric functions of the roots.

Vandermonde now asks whether the general equation of degree n can be solved by a similar expression

$$\frac{1}{n}[(x_1 + \ldots + x_n) + \sqrt[n]{(\rho_1 x_1 + \ldots + \rho_n x_n)^n} + \sqrt[n]{(\rho_1^2 x_1 + \ldots + \rho_n^2 x_n)^n}$$

$$+ \ldots + \sqrt[n]{(\rho_1^{n-1} x_1 + \ldots + \rho_n^{n-1} x_n)^n}]$$

in which $\rho_1, \ldots, \rho_n$ are the n-th roots of unity.

Today, expressions like

$$\rho_1 x_1 + \ldots + \rho_n x_n$$

are called "Lagrange resolvents". Lagrange introduced the same expressions in a memoir to the Berlin Academy in 1771, as we shall see presently. The memoir of Vandermonde, in which the same expressions occur already, was presented to the Paris academy already in 1770, but it was published only in 1774.

In the case of the cubic equation, the method of Vandermonde and Lagrange leads at once to the solution. If i and j are the primitive third roots of unity, one has

$$(x_1 + ix_2 + jx_3)^3 = S + 3iX + 3jY$$

with

$$S = x_1^3 + x_2^3 + x_3^3 + 6x_1 x_2 x_3$$
$$X = x_1^2 x_2 + x_2^2 x_3 + x_3^2 x_1$$
$$Y = x_1 x_2^2 + x_2 x_3^2 + x_3 x_1^2.$$

Here S and $X + Y$ and XY are symmetric functions of the roots, so X and Y are the roots of a quadratic equation. Next, expressions like $x_1 + ix_2 + jx_3$ can be obtained as cube roots, and x_1, x_2, x_3 can be found.

For the biquadratic equation, Vandermonde modifies his approach a little. For degrees larger than 4, his method does not yield a general solution, but in special cases it works. Thus, Vandermonde succeeds in solving the equation

$$x^{11} - 1 = 0.$$

He first reduces it to an equation of degree 5 having the roots

$$\rho + \rho^{-1}, \; \rho^2 + \rho^{-2}, \; \rho^3 + \rho^{-3}, \; \rho^4 + \rho^{-4}$$

where ρ is a primitive eleventh root of unity. Next he solves this quintic equation by introducing his "Lagrange resolvents". These resolvents may be written as

$$L = x_1 + \alpha x_2 + \alpha^2 x_3 + \alpha^3 x_4 + \alpha^4 x_5$$

where α is a primitive fifth root of unity, while $x_1, \ldots, x_5$ are the roots of the quintic equation. In order that L^5 may be expressed rationally, the roots x_i must be taken in a definite order. In the special case of the eleventh roots of unity this order can be found by trial and error, but for the general case a proof is needed. As we shall see in the chapter on Gauss, the essential requirement is, to prove that for every prime number p a "primitive root" g mod p exists such that all integers not divisible by p are congruent to powers of g. The expression "primitive root" is due to Euler, and Gauss has first proved its existence.

Vandermonde claims that the solution of the equation

$$x^n - 1 = 0$$

by his method will always be easy ("nous sera toujours facile"). It seems that he did not see the difficulty of establishing an appropriate order of the roots.

In Chapter 5, we shall explain Gauss' solution of the problem.

Gauss does not quote Vandermonde. Why not? In his paper on Vandermonde quoted right at the beginning of the present section, Lebesgue has discussed this question. He first notes that Gauss has seen the Histoire de l'Académie for the year 1771, in which Vandermonde's memoir was published. According to Lebesgue, Gauss knew the work of Vandermonde. Lebesgue quotes an entry in a notebook of Gauss, in which a paper of Vandermonde on topological questions is mentioned, which appeared in the same volume as Vandermonde's paper on algebraic equations. Lebesgue assumes that Gauss was influenced by Vandermonde's work on the cyclotomic equation. Now why did he not quote Vandermonde in his "Disquisitiones arithmeticae"? Lebesgue's explanation is: Vandermonde did not prove his assertion, and Gauss regarded any mathematical assertion as valuable only if it is accompanied by a full proof. Lebesgue judges this rigorous point of view "profondement injuste". In his opinion, no discovery in mathematics has ever been made by deductive logic. Discoveries always result from "un travail de création de l'imagination": the rigorous proof comes afterwards.

Lagrange

Joseph Louis Lagrange was of Italian origin: he was born in Torino in January 1736. In 1771 he presented an extremely interesting memoir to the Berlin Academy: "Réflections sur la résolution algébrique des équations". It covers 217 pages in Volume 3 of the "Oeuvres de Lagrange" (published by J.-A. Serret in 1869). I shall now summarize the most interesting parts of this Memoir.

Lagrange first considers a cubic equation, which can be written as

$$(1) \qquad\qquad x^3 + nx + p = 0.$$

The solution is well known from the "Ars magna" of Cardano. It can be written in the form

$$(2) \qquad\qquad x = r + s,$$

where r^3 and s^3 are the roots of a quadratic equation. Lagrange shows that r and s can be expressed as functions of the three roots a, b, c of the equation (1):

$$(3) \qquad\qquad r = 1/3(a + \alpha b + \alpha^2 c)$$

$$(4) \qquad\qquad s = 1/3(a + \alpha^2 b + \alpha c),$$

where α is a primitive third root of unity:

(5) $$\alpha^2 + \alpha + 1 = 0.$$

The same results can also be obtained by a direct method, says Lagrange. He starts with an arbitrary linear function of the roots a, b, c

(6) $$y = Aa + Bb + Cc.$$

By permuting the roots one obtains a set of six expressions, which are the roots of an equation of degree 6. If it is required that this equation contains only powers of y^3, it can be shown that the coefficients A, B, C are necessarily proportional to $1, \alpha, \alpha^2$ or to $1, \alpha^2, \alpha$. Thus one obtains, once more, the expressions (3) and (4).

Several fundamental ideas of Galois theory are already present in this part of the treatise of Lagrange. First, the idea that one should express intermediate quantities (like r and s) as *rational functions of the roots* a, b, c. Secondly, Lagrange shows that it is useful to study the behaviour of rational functions like (6) under permutations of the roots. Finally: expressions like (3) and (4), formed by means of roots of unity and called "Lagrange Resolvents", are very useful. As we have seen, the same expressions had already been introduced by Vandermonde in 1770.

Next Lagrange considers a biquadratic equation, which can be written as

(7) $$x^4 + nx^2 + px + q = 0.$$

Lodovico Ferrari has shown that the solution can be obtained by first solving a cubic equation

(8) $$y^3 - \tfrac{1}{2}ny^2 - qy + \tfrac{1}{8}(4nq - p^2) = 0.$$

Lagrange shows that the roots of this equation can be expressed as functions of the roots a, b, c, d of the original equation:

(9) $$u = \tfrac{1}{2}(ab + cd), \quad v = \tfrac{1}{2}(ac + bd), \quad w = \tfrac{1}{2}(ad + bc).$$

If the roots are permuted, the function u gives rise to only three functions u, v, w. This explains why u is a root of a cubic equation.

Very interesting from the historical point of view is Section 100. In this section Lagrange considers rational functions $f(x', x'', \ldots, x^{(n)})$ of the roots of a general equation of degree n. The roots are considered as independent variables or, as we say today, "indeterminates". Lagrange himself once uses the expression "indéterminée".

Two functions t and y of the roots are called *similar* (semblable) if all permutations of the roots that leave t invariant also leave y invariant, and conversely. Lagrange now proves a theorem which will be quoted as "Theorem 100", namely:

If all permutations which leave t invariant also leave y invariant, then y can be expressed as a rational function of t and the coefficients of the given equation.

The idea of the proof is as follows. Let $t', t'', \ldots, t^{(\pi)}$ be the different values the function t assumes when the roots are permuted, and let $y', y'', \ldots, y^{(\pi)}$ be the corresponding values of y. The number π is a divisor of $n!$, and t satisfies an equation $\theta = 0$ of degree π. Hence any function T of t can be written as

$$(10) \qquad T = N_0 + N_1 t + N_2 t^2 + \ldots + N_{\pi-1} t^{\pi-1}.$$

Let $T', T'', \ldots$ be the values of T corresponding to the values $t', t'', \ldots$ of t. Lagrange forms the sum $T'y' + T''y'' + \ldots$ and expresses it as a linear function of the indeterminates $N_0, \ldots, N_{\pi-1}$:

$$(11) \qquad T'y' + T''y'' + \ldots = M_0 N_0 + M_1 N_1 + \ldots + M_{\pi-1} N_{\pi-1}.$$

Now in order to compute y' one has only to specify the coefficients $N_0, N_1, \ldots, N_{\pi-1}$ in such a way that $T'', T''', \ldots$ all become zero. This means that the polynomial (10) is required to have the roots $t'', t''', \ldots$. If this polynomial is multiplied by $t - t'$, it has to be divisible by the polynomial θ. Thus one obtains

$$(12) \qquad (t - t') T = c \theta$$

with constant c. This condition gives rise to a set of linear equations for $N_0, N_1, \ldots, N_{\pi-1}$, which can be solved by elementary calculations, provided the polynomial θ has no double roots. Lagrange chooses $N_0 = 1$, but this is not always possible, and it is not essential for his proof.

Lagrange applies his theorem to equations of degrees 2, 3, and 4. About the quintic he says:

"Il serait à propos d'en faire l'application aux equations du cinquième degré et des degrés supérieurs, dont la résolution est á présent inconnue; mais cette application demande un trop grand nombre de recherches et de combinaisons, dont le succès est encore d'ailleurs fort douteux."

Lagrange also considers special equations such as the "cyclotomic equation"

$$x^n - 1 = 0$$

but he does not go very far. The complete solution of this equation by means of radicals was given by Gauss in 1801 in his admirable "Disquisitiones arithmeticae".

Malfatti

Lagrange presented his "Réflections" to the Berlin Academy in 1771. One year earlier, in 1770, Gianfrancesco Malfatti presented to the Accademia delle Scienze di Siena a highly interesting treatise on quintic equations entitled "De aequationibus quadratocubicis dissertatio analytica". It was published in the Atti della Accademia di Siena in 1771.

According to Raffaela Franci and Laura Toti Rigatelli (Atti del Convegno su G.F. Malfatti, Ferrara, 23-24 ottobre 1981, p. 179-203), Malfatti's treatise is written in bad latin and contains many printing errors, so that it is not easy to read it. However, Franci and Toti Rigatelli have given a very lucid explanation of Malfatti's ideas. The following summary is based on their account.

Malfatti first considers a cubic equation

(1) $$x^3 + 3ax + b = 0.$$

Following Euler, he considers a root x satisfying the linear equation

(2) $$x + m\sqrt[3]{f^2} + n\sqrt[3]{f} = 0.$$

To eliminate the third roots, Malfatti uses a method of Manfredi. Replacing $\sqrt[3]{f}$ by $\alpha\sqrt[3]{f}$ and by $\alpha^2\sqrt[3]{f}$, where α is a third root of unity, and multiplying (2) by the two linear functions of x thus obtained, Malfatti obtains an equation of degree 3:

(3) $$x^3 - 3mnfx + m^3 f^2 + n^3 f = 0.$$

Putting $f = 1$, one gets

(4) $$x^3 - 3mnx + m^3 + n^3 = 0.$$

This equation is equivalent to (1), provided

(5) $$mn = -a$$
$$m^3 + n^3 = b.$$

From this pair of equations one can find m^3 and n^3 and hence m and n. Now Malfatti applies the same method to the quintic equation

(6) $$x^5 + 5ax^3 + 5bx^2 + 5cx + d = 0.$$

He wants to obtain a root x of (6) from the equation

(7) $$x + m\sqrt[5]{f^4} + p\sqrt[5]{f^3} + q\sqrt[5]{f^2} + n\sqrt[5]{f} = 0.$$

Replacing $\sqrt[5]{f}$ by $\alpha\sqrt[5]{f}$, $\alpha^2\sqrt[5]{f}$, $\alpha^3\sqrt[5]{f}$, $\alpha^4\sqrt[5]{f}$, where α is a fifth root of unity, and multiplying (7) by the linear expressions thus obtained, Malfatti obtains a "canonical equation" of degree 5 for x. Putting $f = 1$ and equating the coefficients of the canonical equation with those of (5), Malfatti obtains a set of conditions for m, p, q, n. To simplify these conditions he puts

$$mn = y, \qquad pq = u$$

and

$$25uy - 5a^2 + 5c/3 = z.$$

After tedious calculations Malfatti ends up with an equation of degree 6 for z. In the general case this equation has no rational factor of degree 1 or 2 or 3, but if it has, the given equation (6) can be solved by radicals. On can then determine first z and next m, p, q, n, and one finally obtains the roots:

$$
\begin{aligned}
x_0 &= -(m+p+q+n) \\
x_1 &= -(\alpha m + \alpha^2 p + \alpha^3 q + \alpha^4 n) \\
x_2 &= -(\alpha^2 m + \alpha^4 p + \alpha q + \alpha^3 n) \\
x_3 &= -(\alpha^3 m + \alpha p + \alpha^4 q + \alpha^2 n) \\
x_4 &= -(\alpha^4 m + \alpha^3 p + \alpha^2 q + \alpha n).
\end{aligned}
$$

(8)

It is easy to solve the linear equations (8) for m, p, q, n. It follows that m, p, q, n are linear functions of the roots, and that z is a biquadratic function of the roots.

Independent of Malfatti, Lagrange too constructed a "resolvent" z, which is a function of the roots assuming six values when the roots are permuted. This function is "similar" to Malfatti's z in the sense of Lagrange, for both are invariant under a group of 20 permutations of the roots x_k. In modern notation the permutations of this group can be defined by the formulae

$$
(9) \qquad\qquad k' \equiv gk + h \qquad (\mathrm{mod}\ 5)
$$

with $g = 1, 2, 3$ or 4, and $h = 0, 1, 2, 3$ or 4. We shall see that the same group plays an important role in the theory of Galois.

Ruffini

Paolo Ruffini, born 1765, was a student and admirer of Lagrange. He published several papers claiming to prove that the general equation of fifth degree or higher cannot be solved by radicals. His first treatise, published in 1898 in Bologna (Opere matematiche di Paolo Ruffini, Vol. 1, p. 1–324), is entitled

"Teoria generale delle equazioni, in cui si dimostra impossibile la soluzione algebraica delle equazioni generale di grado superiore al quarto",
which means:

General theory of equations, in which it is proved that the algebraic solution of the general equations of degree larger than four is impossible.

A few years later Ruffini wrote a small treatise entitled "Rischiarimenti e risposte alle obbiezioni" (Elucidations and Answers to Objections), which was found among his manuscripts and printed in Opere matematiche, Vol. 1, p. 327–342.

A second memoir on the solution of equations of degrees larger than 4 was published in Volume 9 of the Memorie di Matematica e Fisica della Società Italiana delle Scienze (Modena 1802), and republished in Ruffini's Opere I, p. 345–406.

Finally, in 1813, Ruffini published a third memoir entitled "Riflessioni intorno alla soluzione delle equazioni algebraiche generali" (Opere II, p. 155–268), in which the ideas of his earlier memoirs were further elaborated.

Ruffini's methods were essentially those of Lagrange. He considered rational functions of the roots of a general equation of degree n. If p is the number of permutations that leave such a function unaltered, p is a divisor of $n!$, and the number of different values the function assumes if the roots are permuted is $n!/p$. As Lagrange had already shown, such a function is a root of an equation of degree $n!/p$. Ruffini studies in great detail the set of p permutations leaving the function unaltered. In particular, he shows that in the case of the quintic the degree $n!/p$ can be 2 or 5 or 6, but not 3 or 4, which means that a resolvent in the sense of Lagrange satisfying an equation of degree 3 or 4 is impossible. If $n!/p$ is not 2, it must be divisible by 5. If $n!/p$ is 5, resolvents of degree 5 exist, but they cannot be reduced to binomial equations

$$z^5 - m = 0.$$

Ruffini claims to have proved that the general quintic equation cannot be solved by radicals, but his proof is not conclusive, because it is based on the hypothesis that these radicals can all be expressed as rational functions of the roots. It was Abel who first completed the proof by showing that the radicals needed for solving an equation can always be chosen as rational functions of the roots of the equation and of certain roots of unity.

Ruffini's proof was not well received by his contemporaries and successors. Malfatti criticized Ruffini's proof and concluded that there still remain doubts whether the general solution of the quintic equation is impossible. Malfatti's note, entitled "Dubbii proposti al socio Paolo Ruffini sulla sua dimostrazione della impossibilità di risolvere le equazioni superiori al quarto grado", was published in Vol. 11 of the Memorie di Matematica e Fisica della Società Italiana delle Scienze, p. 579–607.

Carnot and Legrendre expressed similar doubts. See p. 59 of B.M. Kiernan: Galois Theory from Lagrange to Artin, Archive for History of Exact Sciences 8, p. 40–154 (1971).

Cauchy seems to have considered Ruffini's proof as conclusive (see p. 60 of Kiernan's paper), but Abel expressed his doubts thus:

Le premier, et, si je ne me trompe, le seul qui avant moi ait cherché à démontrer l'impossibilité de la résolution algébrique des équations générales, est le géomètre Ruffini; mais son mémoire est tellement compliqué qu'il est très difficile de juger de la justesse de son raisonnement. Il me paraît que son raisonnement n'est pas toujours satisfaisant.

A thorough study of Ruffini's work and its relation to the theory of substitution groups is due to Heinrich Burkhardt. It was published in 1892 under the title "Die Anfänge der Gruppentheorie und Paolo Ruffini" in Zeitschrift für Mathematik und Physik 37, Supplement (Abhandlungen zur Geschichte der Mathematik VI), p. 121–159. Burkhardt shows that several fundamental notions of the theory of permutation groups are already present in the work of Ruffini. In particular, Ruffini distinguishes between what we now call transitive and intransitive groups.

Cauchy

Ruffini had proved that the number of different values which a non-symmetric rational function attains cannot be less than 5, unless it is 2. Cauchy generalized this result of Ruffini to functions of n variables. His paper of 1815 containing this generalization is entitled "Sur le Nombre des Valeurs qu'une Fonction peut acquérir, lorsqu'on y permute de tous les manières possible les quantités qu'elle renferme", Journal de l'Ecole Polytechnique, cahier 17, tome 10, p. 1–28 (Oeuvres complètes d'Augustin Cauchy, 2me série, Vol. 1, p. 64–90).

Let n be the number of independent variables, and p the largest prime number contained in n. Cauchy proved: The number of different values of a non-symmetric rational function of n variables cannot be less than p, unless it is 2.

Cauchy introduced a distinction between permutations and substitutions. If one writes the n variables in any order, one has a *permutation*. A *substitution* is the passage from one permutation to another. The passage from the permutation 1.2.4.3 to 2.4.3.1 is denoted by

$$\begin{pmatrix} 1 . 2 . 4 . 3 \\ 2 . 4 . 3 . 1 \end{pmatrix}.$$

Galois used the same terminology, although not consistently. He defined the notion "group of substitutions". In later times, the "substitutions" of Cauchy and Galois were often called "permutations", in agreement with the original meaning of the Latin verb permutare.

Cauchy also defined products of substitutions. The product ST is obtained by first performing S and next T. The same convention was used by Galois.

During the years 1844–1846 Cauchy published a sequence of papers on substitutions. He calls two substitutions "similar", if they partition into cycles in the same way. He proves that P and Q are similar if and only if Q is equal to $R^{-1}PR$. He also proves that the order of any group of substitutions is divisible by the order of any substitution in the group, and that the order of any group of substitutions on n variables is a divisor of $n!$. The latter theorem had already been proved by Lagrange. Cauchy's proof is the same as that of Lagrange: the group of all substitutions is partitioned into cosets of the subgroup. The same method was used later on by Jordan in order to prove that the order of any subgroup of a finite group divides the order of the group. Jordan generously attributed this theorem to Lagrange and Cauchy.

Abel

Still in his teens, the extremely gifted young mathematicien Niels Henrik Abel (born 1802) thought that he could solve the general quintic equation, but he soon discovered his error. In the spring of 1824 he succeeded in proving that a solution by radicals is impossible. At his own expense he published a pamphlet in French entitled "Mémoire sur les équations algébriques" (Oeuvres complètes de Niels Henrik Abel, publiées par L. Sylow et S. Lie, Vol. 1, p. 28–

33), in which he presented a completely clear proof of this impossibility. A new, more elaborated version of the same proof was published in 1826 in Crelle's Journal für die reine und angewandte Mathematik, Vol. 1, under the title "Démonstration de l'impossibilité de la résolution algébrique des équations générales qui passent le quatriéme degré (Oeuvres I, p. 66–87). The main ideas are the same in both papers, but some parts were expanded and other parts simplified in the later paper.

Abel makes use of the results obtained by Lagrange and Cauchy concerning the number of values a function of n variables can attain if the variables are permuted. However, the essential point in his proof is the first step, which is quite new.

Abel starts with an equation

$$(1) \qquad y^5 - a y^4 + b y^3 - c y^2 + d y - e = 0$$

in which the coefficients are "general", that is, they are just letters or independent variables. Supposing that one can express y as a function of the coefficients by means of radicals, Abel states that one can write y as

$$(2) \qquad y = p + p_1 R^{1/m} + p_2 R^{2/m} + \ldots + p_{m-1} R^{(m-1)/m}$$

where m is a prime number. The quantities $R, p, p_1, \ldots, p_{m-1}$ are supposed to be expressions of the same form as y, involving other radicals, and so on until one arrives at rational functions of the coefficients of the original equation. In the terminology of Galois, one starts with the field of rational functions of a, b, c, d, e with constant coefficients, and one "adjoins" one radical with prime exponent after the other. Among the constant coefficients Abel always includes the m-th roots of unity, where m is any one of the prime exponents used in the solution.

One can suppose, says Abel, that $R^{1/m}$ cannot be expressed as a rational function of $a, b, \ldots, p, p_1, p_2, \ldots$ for otherwise the adjunction of the radical would be superfluous. One can also suppose that in (2) not all coefficients $p_1, p_2, \ldots$ are zero.

In his first paper Abel supposes $p_1 \neq 0$ (in his second paper he shows that this restriction is not essential). Now replacing R by R/p_1^m, one can make p_1 equal to 1. Putting $R^{1/m} = z$, one has

$$(3) \qquad y = p + z + p_2 z^2 + \ldots + p_{m-1} z^{m-1}.$$

Substituting this value into (1), one obtains a result of the form

$$(4) \qquad q + q_1 z + q_2 z^2 + \ldots + q_{m-1} z^{m-1} = 0$$

in which $q, q_1, q_2, \ldots$ are polynomials in $a, b, \ldots p_1, p_2, \ldots$ and R.

Now comes the crucial step. Abel asserts: For (4) to be valid, it is necessary that

$$q = 0, \; q_1 = 0, \ldots, q_{m-1} = 0.$$

The proof is very ingenious. The two Equations (4) and

(5) $$z^m - R = 0$$

have a root z in common. If $q, q_1, \ldots$ are not zero, the number of roots they have in common is at most $m-1$. Let k be this number. Then, by calculating the largest common divisor of the polynomials on the left of (4) and (5), one can find an equation of degree k

(6) $$r + r_1 z + r_2 z^2 + \ldots + r_k z^k = 0.$$

For the continuation of the proof I shall follow the simplified exposition in Abel's second paper. If the polynomial on the left of (6) is factorized, one of the factors must be zero. Thus, one obtains an irreducible equation for z:

(7) $$t_0 + t_1 z + \ldots + t_{\mu-1} z^{\mu-1} + z^\mu = 0.$$

One can suppose, says Abel, that it is impossible to find an equation of the same form of lower degree. This equation has its μ roots in common with the equation (5). Now all roots of this latter equation are of the form αz. The degree μ is at least 2, for otherwise z would be a rational function of $a, b, p, p_1, p_2, \ldots$. It follows that the equation (7) has at least two roots z and αz:

(8)
$$t_0 + t_1 z + t_2 z^2 + \ldots + t_{\mu-1} z^{\mu-1} + z^\mu = 0$$
$$t_0 + \alpha t_1 z + \alpha^2 t_2 z^2 + \ldots + \alpha^{\mu-1} t_{\mu-1} z^{\mu-1} + \alpha^\mu z^\mu = 0.$$

Multiplying the first equation by α^μ and subtracting it from the second, one obtains an equation of degree less than μ, which is impossible. Hence in (4) all coefficients $q, q_1, \ldots, q_{m-1}$ must be zero.

The equation (4) was obtained by substituting (3) into (1) and using (5). Now (5) is satisfied not only by z but also by

$$\alpha z, \alpha^2 z, \ldots, \alpha^{m-1} z.$$

Hence, replacing $R^{1/m}$ in (2) by $\alpha^k R^{1/m}, \ldots$, one always obtains roots of the equation (1). These roots are all different, hence m cannot be larger than 5, and if the roots thus obtained are called $y_1, \ldots, y_m$ one has

$$y_1 = p + z + p_2 z^2 + \ldots + p_{m-1} z^{m-1}$$
$$y_2 = p + \alpha z + \alpha^2 p_2 z^2 + \ldots + \alpha^{m-1} p_{m-1} z^{m-1}$$
$$\ldots$$
$$y_m = p + \alpha^{m-1} z + \alpha^{m-2} p_2 z^2 + \ldots + \alpha p_{m-1} z^{m-1}.$$

These equations can easily be solved for $p, z, p_2 z^2, \ldots, p_{m-1} z^{m-1}$. It follows that $p, p_2, \ldots, p_{m-1}$ and $z = R^{1/m}$ are rational functions of the roots $y_1, \ldots, y_5$ of the equation (1). Of course, $R = z^m$ is also a rational function of the roots.

The quantity R may be given as a rational function of an earlier radical $v^{1/n}$. This function can be written as

$$(9) \qquad R = S + v^{1/n} + S_2 v^{2/n} + \ldots + S_{n-1} v^{(n-1)/n}.$$

If this quantity is treated in the same way as the y of equation (2), one sees that either the adjunction of $v^{1/n}$ is unnecessary, or the quantities $v^{1/n}, S, S_2, \ldots$ can be expressed as rational functions of the roots $y_1, \ldots, y_5$. By repeating the same reasoning one concludes that all irrational quantities occurring in the expression of the roots y are rational functions of these roots.

This is just the hypothesis from which Ruffini started in his proof of the unsolvability of the quintic equation. This hypothesis is now fully justified.

From this point onwards, Abel was able to use the methods and results of Lagrange, Ruffini, and Cauchy. In particular, he used (and quoted) the result of Cauchy which says that the number of values a rational function can attain cannot be 3 or 4, which implies that the number m can only be 2 or 5. Abel discusses the two cases $m = 5$ and $m = 2$ separately, and he concludes that in both cases the solution of the general quintic equation by radicals is impossible.

Two months before his death in 1829, Abel published another paper, entitled "Mémoire sur une classe particulière d'équations résoluble algébriquement", Crelle's Journal für die reine und angewandte Mathematik, Vol. 4 (Oeuvres I, p. 478-514).

This memoir deals with a particular class of equations of all degrees which are solvable by radicals. To this class belongs the cyclotomic equation $x^n - 1 = 0$. Abel proves the following general theorem:

If the roots of an equation are such that all roots can be expressed as rational functions of one of them, say x, and if any two of the roots, say θx and $\theta_1 x$ (where θ and θ_1 are rational functions) are connected in such a way that

$$(10) \qquad \theta \theta_1 x = \theta_1 \theta x,$$

then the equation can be solved by radicals.

Today, groups in which multiplication is commutative are called Abelian, and equations having the property (10) are called, since Kronecker (1853), *Abelian equations*.

Abel's theorem just stated is a special case of a main theorem of Galois theory, which says:

An equation is solvable by radicals if and only if its Galois group G is solvable, that is, if G possesses a composition series

$$G \supset H_1 \supset H_2 \supset \ldots \supset H_m = E$$

in which all indices are prime numbers. It is easy to see that every finite Abelian group is solvable. Galois presented a proof of his main theorem to the Academy of Paris in May 1829, in the same year in which Abel's paper was published.

Chapter 5
Carl Friedrich Gauss

The most important contributions of Gauss to the theory of algebraic equations are:

first, the complete solution of the "cyclotomic equation"

(1) $$x^m - 1 = 0$$

by means of radicals,

second, the proof that every polynomial in one variable with real coefficients is a product of linear and quadratic factors. This theorem implies what we now call the "fundamental theorem of algebra": Every polynomial $f(x)$ with complex coefficients is a product of linear factors.

We shall now discuss these two extremely interesting contributions.

The Cyclotomic Equation

The equation (1) is called cyclotomic, because its solution is closely connected with the construction of a regular polygon of n sides inscribed in a given circle.

To see this, one has only to note that the equation (1) has n complex roots

(2) $$\cos(2\pi k/n) + i\sin(2\pi k/n) \quad k = 0, 1, 2, \ldots, n-1.$$

This trigonometric solution was known to De Moivre and Euler long before Gauss. Now if one represents the complex numbers $a + ib$ by points in the plane with orthogonal coordinates (a, b), it is clear that the complex numbers (2) are represented by the vertices of a regular n-gon inscribed in the unit circle. Hence, if one succeeds in solving the equation (1) by means of square roots, one can construct the regular n-gon with ruler and compass.

The Pythagoreans already knew how to construct regular polygons of 3, 4, 5, and 6 sides. Their constructions can be found in Book 4 of the Elements of Euclid. For the ascription of this book to the Pythagoreans see my book "Die Pythagoreer" (Artemis-Verlag, Zürich 1979), p. 348–351.

Lagrange solved the equation

(3) $$x^5 - 1 = 0$$

as follows. One root is $x=1$. The others are roots of the equation

$$x^4+x^3+x^2+x+1=0,$$

which can be written as

(4) $$(x^2+x^{-2})+(x+x^{-1})+1=0.$$

   Putting

(5) $$x+x^{-1}=y$$

one obtains

(6) $$y^2+y-1=0.$$

   This quadratic equation can be solved for y, and next (5) can be solved for x. It follows, once more, that the regular pentagon can be constructed by means of ruler and compass.

   Euclid's construction is also based on the solution of a quadratic equation. Proposition 11 in Book 2 of Euclid's Elements reads in the translation of Heath:

   To cut a given straight line so that the rectangle contained by the whole and one of the segments is equal to the square on the remaining segment.

   If the given straight line is called a and the second segment y, Euclid's problem is, to solve the equation

(7) $$a(a-y)=y^2.$$

   In his solution of the problem II, 11 Euclid first solves the equivalent equation

(8) $$y^2+ay=a^2$$

and next he subtracts the rectangle ay on both sides, thus obtaining the solution of (7). If the given segment a is taken as a unit of length, it is seen that (8) is the same as Lagrange's equation (6).

   In Book 4, Euclid uses the solution of (7) in his construction of the regular pentagon. Just so, Lagrange uses the solution of (6) for the solution of the cyclotomic equation (3).

   Lagrange next applies the same method to the equation

(9) $$x^{11}-1=0$$

(Oeuvres III, p. 246). Dividing by $x-1$ and next by x^5, Lagrange obtains

(10) $$(x^5+x^{-5})+(x^4+x^{-4})+(x^3+x^{-3})+(x^2+x^{-2})+(x+x^{-1})+1=0.$$

Putting again

$$x + x^{-1} = y$$

one obtains a quintic equation for y.

Lagrange left it at that, but Vandermonde succeeded in solving the quintic equation by radicals, as we have seen in Chapter 4.

At the age of not quite 19 years, Gauss discovered that the regular 17-gon can be constructed with ruler and compass. In Chapter 7 of the famous work of Gauss entitled "Disquisitiones arithmeticae" the full proof of the solvability of the equation (1) by radicals was given. The equation

(12) $$x^{17} - 1 = 0$$

is treated as a special case. Since we do not know how the young Gauss found the solution of (12) and hence the construction of the 17-gon, we have no other choice than to follow Gauss and to treat the general case first.

Gauss first shows that the general equation (1) can be reduced to the special case in which n is prime, by writing n as a product of powers of primes. A special case, namely $n = 15$, was already known to Euclid. Euclid shows: if one can inscribe in a circle a regular triangle and a regular pentagon, one can also inscribe a regular polygon of 15 sides.

Dividing (1) by $x - 1$, one obtains the equation

(13) $$X = x^{n-1} + x^{n-2} + \ldots + x + 1 = 0.$$

Supposing n to be prime, Gauss first proves that the polynomial X is *rationally irreducible*. Next he announces his main result: If $n - 1$ is a product of factors $\alpha \beta \gamma \ldots$, the equation (1) can be solved by solving equations of degrees $\alpha, \beta, \gamma, \ldots$. For instance, if n is 17, we have

$$n - 1 = 2^4,$$

so the equation (12) can be solved by solving four quadratic equations, and hence the 17-gon can be constructed with ruler and compass. More generally, if $n - 1$ is a power of 2, which happens for

(14) $$n = 3, 5, 17, 257, 65537,$$

the regular n-gon can be constructed with ruler and compass.

The primes mentioned in (14) were known to Gauss. Other primes of the form $2^m + 1$ are not known up to the present day (December 3, 1982).

Still supposing n to be prime, Gauss denotes by r any one of the roots of (13). Now the roots are

(15) $$r, r^2, \ldots, r^{n-1}.$$

Two powers r^λ and r^μ are multiplied by adding the exponents and reducing the sum $\lambda + \mu$ modulo n.

Gauss next notes that every rational function of the roots can be written as

(16) $$A + A'r + A''r^2 + \ldots + A^{(n-1)}r^{n-1}.$$

To simplify the notation, Gauss writes $[\lambda]$ instead of r^λ. Thus, the roots (15) are rewritten as

(17) $$[1], [2], \ldots, [n-1].$$

In Chapter III of the Disquisitiones, Gauss has proved: if n is prime, the multiplicative group of integers modulo n is cyclic, i.e. a "primitive element" g exists such that all exponents not divisible by n are congruent to powers of g. So the roots (17) can be reordered and written as

(18) $$[1], [g], [g^2], \ldots, [g^{n-2}].$$

This reordering is an essential point in the theory of Gauss. The exponents of g are called *indices*. They play the role of logarithms: two powers of g are multiplied by adding their indices (mod $n-1$).

Now let e be any divisor of $n-1$. Putting

$$n - 1 = ef$$

and

$$g^e = h,$$

Gauss considers the set of roots

$$[\lambda], [\lambda h], [\lambda h^2], \ldots, [\lambda h^{f-1}],$$

where λ is an arbitrary integer incongruent to zero (mod n), and he forms the sum

(19) $$(f, \lambda) = [\lambda] + [\lambda h] + [\lambda h^2] + \ldots + [\lambda h^{f-1}].$$

These sums are independent of the choice of g. They are called *periods*.

Gauss elucidates the formation of the periods by working out the example $n = 19$. I prefer to give the example $n = 17$, elaborated by Gauss in Section 354 (Werke, Vol. I, p. 437). As a primitive element (mod 17) Gauss chooses $g = 3$. Thus, the indices (mod 16)

$$i = 0, 1, 2, 3, 4, 5, 6, 7, 8, 9, 10, 11, 12, 13, 14, 15$$

give rise to the powers of 3 (mod 17)

$$\mu = g^i = 1, 3, 9, 10, 13, 5, 15, 11, 16, 14, 8, 7, 4, 12, 2, 6$$

and to the roots

$$[\mu] = r^\mu = r, r^3, r^9, r^{10}, \ldots, r^6.$$

The divisors of $n-1=16$ are

$$e=1, 2, 4, 8, 16$$

corresponding to

$$f=16, 8, 4, 2, 1.$$

There is only one period $(16, 1)$, namely the sum of all roots. There are two periods with $f=8$, namely

and

$$(8, 1)=[1]+[9]+[13]+[15]+[16]+[8]+[4]+[2]$$

$$(8, 3)=[3]+[10]+[5]+[11]+[14]+[7]+[12]+[6].$$

There are four periods with $f=4$, namely

$$(4, 1), (4, 3), (4, 9), (4, 10).$$

There are eight periods with $f=2$, such as

$$(2, 1)=[1]+[16]=r+r^{-1},$$

and there are 16 periods with $f=16$, namely the single roots.

Gauss also considers the period $(f, 0)$, which is a sum of f units and hence equal to f.

In Section 345 Gauss proves a general theorem to the effect that a product

$$(f, \lambda) \cdot (f, \mu)$$

can be expressed as a sum of periods thus:

(20) $$(f, \lambda) \cdot (f, \mu)=(f, \lambda+\mu)+(f, \lambda'+\mu)+(f, \lambda''+\mu)+\dots.$$

Now let us apply formula (20) to the case $n=17$. The sum

$$(8, 1)+(8, 3)$$

is the sum of all roots and hence equal to -1. The product

$$(8, 1) \cdot (8, 3)$$

can be computed by (20): it is -4. Hence $(8, 1)$ and $(8, 3)$ are the roots of the quadratic equation

(21) $$y^2+y-4=0.$$

By solving this equation, one obtains $(8, 1)$ and $(8, 3)$. Next $(4, 1)$ and $(4, 9)$ can be computed by the same method. Their sum is $(8, 1)$ and their product -1, so they are the roots of the quadratic equation

(22) $$x^2 - (8, 1)x - 1 = 0.$$

Just so, $(4, 3)$ and $(4, 10)$ are the roots of the equation

(23) $$x^2 - (8, 3)x - 1 = 0.$$

By the same method the periods $(2, \lambda)$ and finally the roots $[\mu]$ can be obtained as roots of quadratic equations.

In the general case, one has to factorize $n - 1$

$$n - 1 = \alpha\beta\gamma\ldots$$

and to solve equations of degrees $\alpha, \beta, \gamma, \ldots$. In § 359, Gauss shows that these equations can be solved by radicals.

I suppose that these examples are sufficient to explain the main ideas of Gauss on the subject of the cyclotomic equation.

The "Fundamental Theorem"

In the notation of Gauss, every algebraic equation of degree m can be written as

(24) $$x^m + Ax^{m-1} + Bx^{m-2} + \ldots + M = 0$$

or $X = 0$. The so-called "Fundamental Theorem of Algebra" says that every polynomial X with real or complex coefficients can be factored into linear factors in the field of complex numbers.

It is sufficient to prove the theorem for polynomials with real coefficients, for if X has complex coefficients, the product $X\bar{X}$ is real, and its factorization implies the factorization of the factors X and $\bar{X}$. So Gauss is justified in restricting himself to real polynomials X.

In his first proof Gauss does not introduce complex numbers. He proves the fundamental theorem in the following form:

Every polynomial X with real coefficients can be factored into linear and quadratic factors.

Gauss considered the theorem so important that he has given four proofs. The principles, on which the first proof is based, were discovered by Gauss in October 1797. The first proof was published in 1799, the second and third in 1816, and the fourth in 1849. The fourth proof is based on the same principles as the first. Here I shall restrict myself to the first three proofs.

All four proofs have been translated from Latin into German by E. Netto and published under the title "Die vier Gauss'schen Beweise für die Zerlegung ganzer algebraischer Funktionen in reelle Faktoren ersten oder zweiten Grades", Ostwald's Klassiker der exakten Wissenschaften, Vol. 14 (Leipzig 1913).

The First Proof

The first proof of Gauss was published in his dissertation (Werke III, p. 1–30). Before exposing his own proof, Gauss critizes earlier proofs given by d'Alembert, Euler, Fontenex, and Lagrange. His main objection is that in all these proofs the existence of roots is presupposed. It is shown that the roots can be obtained as complex numbers, provided they exist in some sense or other. There are other objections to the single proofs, which will not be discussed here.

Gauss starts with a real polynomial

$$(25) \qquad X = x^m + A x^{m-1} + B x^{m-2} + \ldots + L x + M,$$

in which x is an indeterminate ("unbestimmte Größe"). What he wants to prove is that a linear or quadratic factor of X exists. A real linear factor implies the existence of a real root $\pm r$, where r is positive or zero. An irreducible quadratic factor implies the existence of two complex roots

$$(26) \qquad r(\cos \varphi \pm i \sin \varphi),$$

hence the quadratic factors can be written as

$$(27) \qquad x^2 - 2 x r \cos \varphi + r^2 \quad (r > 0).$$

Substituting one of the roots (26) into the equation $X = 0$ and separating the real and imaginary parts, one obtains a pair of real equations for r and φ:

$$(28) \qquad r^m \cos m \varphi + A r^{m-1} \cos(m-1) \varphi + \ldots + L r \cos \varphi + M = 0$$

$$(29) \qquad r^m \sin m \varphi + A r^{m-1} \sin(m-1) \varphi + \ldots + L r \sin \varphi = 0.$$

Gauss notes that Euler obtained this pair of equations by using complex numbers. Gauss avoids complex numbers: he derives (28) and (29) directly from the assumption that the polynomial X has a linear factor $x \pm r$ or a quadratic factor (27).

Gauss interprets (28) and (29) as equations of algebraic curves in polar coordinates, and he proceeds to prove that these curves intersect in at least one point. If this is proved, it follows that X has a linear or quadratic factor, and by continuing the process one obtains a factorization of X into linear and quadratic factors.

The equation (28) is called $U=0$, and (29) is called $T=0$. To illustrate the proof, I have drawn the curves $U=0$ and $T=0$ for the case of a quadratic equation

$$x^2+1=0.$$

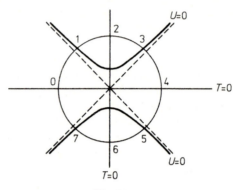

Fig. 23

In orthogonal coordinates x and y we have two curves of order m. The axis $y=0$ is always a part of the second curve $T=0$.

Gauss now studies the intersections of the two curves with a circle of radius R, and he proves:

For a sufficiently large radius R there are exactly $2m$ intersections of the circle with $T=0$ and $2m$ intersections with $U=0$, and every point of intersection of the second kind lies between two points of intersection of the first kind.

Gauss presents a complete proof of this lemma. He next notes that the $4m$ points change only very little if R is made a little larger or smaller. In modern terminology we would say that the $4m$ points are continuous functions of R. Gauss does not prove this continuity: he only says that it is "easy to see". Next Gauss studies the behaviour of the branches of the curves $U=0$ and $T=0$ inside the circle, and he asserts: There exists a point of intersection of a branch of the first curve with a branch of the second curve. For this conclusion he gives an intuitive, geometrical proof. He denotes the point of intersection of the circle with the negative x-axis by 0, the next neighbouring point on the circle by 1, and so on, as in Fig. 23. The odd numbers denote points on $U=0$, the even numbers points on $T=0$. Now he says: If a branch of an algebraic curve enters a certain domain, it must also leave the same domain somewhere. In a footnote he adds:

It seems to be well demonstrated that an algebraic curve neither ends abruptly (as it happens in the transcendental curve $y=1/\log x$), nor does it quasi loose itself after an infinite number of windings in a point (like a logarithmic spiral). As far as I know nobody has ever doubted this, but if anybody requires it, I take it on me to present, on another occasion, an indubitable proof.

If this starting point is accepted, it follows that every "even point" is connected with (at least) another even point by a branch of the curve $T=0$, and every "odd point" with another odd point by a branch of the curve $U = 0$. Now, however complicated these connections may be, one can show that a point of intersection always exists. This is proved as follows.

Suppose that no point of intersection exists. The point 0 is connected with the point $2m$ by the x-axis. The point 1 cannot be connected with any point on the other side of this axis without intersecting the axis. So, if the point 1 is connected with the odd point n, we must have $n < 2m$. Just so, if 2 is connected with n', we must have $n' < n$. Note that the difference $n' - 2$ is even, because 2 and n' are both even. Continuing in this way, one finally finds a point h connected with $h + 2$. But now the branch which enters the circle at the point $h + 1$ must necessarily intersect the branch connecting h with $h + 2$, contrary to our hypothesis. Hence there exists a point of intersection.

From this exposition one sees that the first proof of Gauss is based upon assumptions about the branches of algebraic curves, which appear plausible to our geometrical intuition, but which are not strictly proved by Gauss. Alexander Ostrowski has shown in a very interesting paper "Über den ersten und vierten Gauss'schen Beweis des Fundamentalsatzes der Algebra", that all assumption made by Gauss can be justified by indubitable proofs. Ostrowski's paper was first published in Nachrichten der Gesellschaft der Wissenschaften Göttingen 1920, and reprinted in Gauss' Werke X, 2.

The Second Proof

The second proof is purely algebraic. The only suppositions made about the field of real numbers are:

1° that every real equation of an odd degree has a real root,

2° that every quadratic equation with complex coefficients has two complex roots.

The idea underlying the second proof is simple, but the working out is rather difficult. Gauss starts with a real polynomial of degree m

$$(30) \qquad Y = x^m - L'x^{m-1} + L''x^{m-2} - \ldots + \ldots$$

If one supposes for a moment that Y can be factored into linear factors

$$(31) \qquad Y = (x - a)(x - b)(x - c)\ldots$$

in some field extension, then a linear combination

$$(32) \qquad (a + b)t - ab$$

can be formed with a new indeterminate t. If the roots $a, b, c, \ldots$ are permuted, the linear function (32) assumes

$$m' = \tfrac{1}{2}m(m + 1)$$

values, hence it is a root of an equation of degree m'. The roots of this auxiliary equation are linear functions of t of the form (32). As soon as one root of the auxiliary equation is known, $a+b$ and ab are known, so a and b can be expressed by means of square roots. This remains true if the indeterminate t is specialized in such a way that different linear functions (32) remain different after the specialization.

Now if m is a number of the form

$$(33) \qquad\qquad m = 2^\mu k$$

where k is odd, the degree of the auxiliary equation is of the form

$$(34) \qquad\qquad m' = 2^{\mu-1} k'$$

where k' is again odd.

As soon as a complex root of this auxiliary equation is known, two roots a and b of the original equation can be computed as complex numbers by extracting a square root.

Continuing in the same way, one finally arrives at an equation of an odd degree. The coefficients of this equation are symmetric functions of the roots $a, b, \dots$ with real coefficients, so they are known real numbers. Since the degree is odd, this equation has at least one real root. Going back trough the sequence of auxiliary equations, one can compute at least one complex root of the original equation.

In this simplified form, the proof works if it is known that the equation $Y = 0$ has m roots $a, b, \dots$ in some extension of the field of real numbers. The existence of such an extension can be proved by Kronecker's method of "symbolic adjunction": the proof can be found in any textbook of modern algebra. However, Gauss does not follow this road. He constructs his auxiliary equations without assuming the existence of the roots. For instance, he constructs the auxiliary equation of degree m' as follows.

First, the special polynomial (30) is replaced by a polynomial y, the roots of which are indeterminates $a, b, c, \dots$

$$(35) \qquad\qquad y = (x-a)(x-b)(x-c)\dots .$$

Gauss next forms an auxiliary polynomial ζ in a new variable u, defining ζ as the product of the m' expressions

$$(36) \qquad\qquad u - (a+b)t + ab$$

obtained by permuting the roots. This polynomial ζ is symmetric in the indeterminates $a, b, c, \dots$, hence it can be expressed in a unique way as a polynomial in u and t and the coefficients of y, which are the elementary symmetric functions of $a, b, c, \dots$. After this, the coefficients of y are replaced by the coefficients $L, L', \dots$ of the given polynomial (30), and thus the auxiliary polynomial Z is obtained.

In Sections 12-15 Gauss proves a theorem:

If the discriminant of Y is not zero, the discriminant of Z cannot be zero.

The proof of this theorem covers four pages in Netto's translation. Right at the beginning of the proof Gauss says: "The proof of this theorem would be extremely simple if we could presuppose that Y is a product of linear factors."

Afterwards Gauss substitutes for t such a real value that the discriminant of Z is still different from zero, and he shows: if a root of Z is known, a pair of roots of the original polynomial Y can be computed. Obviously, he first derived his method of finding a root of Y from the assumption that Y is a product of prime factors, and afterwards he reshaped his proofs so as to make them independent of this assumption.

The Third Proof

The third proof of Gauss is much simpler. According to his own testimony he found this proof by continued thinking after the second proof was printed in 1816.

Gauss starts with a polynomial

(37) $$X = z^m + A z^{m-1} + B z^{m-2} + \ldots + L z + M$$

with real coefficients. He puts

(38) $$r^m \cos m\varphi + A r^{m-1} \cos(m-1)\varphi + \ldots + L r \cos \varphi + M = t$$

(39) $$r^m \sin m\varphi + A r^{m-1} \sin(m-1)\varphi + \ldots + L r \sin \varphi = u.$$

The expressions t and u are the same as the expressions U and T in Gauss' first proof. They are the real and imaginary parts of the complex expression obtained by substituting

(40) $$z = r(\cos \varphi + i \sin \varphi)$$

into (37).

The derivatives of t and u with respect to φ are called $-u'$ and $+t'$. Thus we have

$$t' = m r^m \cos m\varphi + (m-1) A r^{m-1} \cos(m-1)\phi + \ldots + L r \cos \varphi$$

$$u' = m r^m \sin m\varphi + (m-1) A r^{m-1} \sin(m-1)\varphi + \ldots + L r \sin \varphi.$$

Gauss now proves that

$$t t' + u u'$$

is positive for a sufficiently large value R of r, no matter what value φ has. It is easy to see that this is true. For large r the main terms of t and u are

$$r^m \cos m\varphi \quad \text{and} \quad r^m \sin m\varphi$$

and the main terms of t' and u' are

$$m\, r^m \cos m\, \varphi \quad \text{and} \quad m\, r^m \sin m\, \varphi.$$

So the main term of $t\, t' + u\, u'$ is

$$m\, r^{2m}(\cos^2 m\, \varphi + \sin^2 m\, \varphi) = m\, r^{2m}$$

which is positive.

The second derivatives of t and u with respect to φ are called $-u''$ and $+t''$:

(41) $$t'' = m^2\, r^m \cos m\, \varphi + \ldots + L\, r \cos \varphi$$

(42) $$u'' = m^2\, r^m \sin m\, \varphi + \ldots + L\, r \sin \varphi.$$

Gauss wants to show that there is a point in the plane at which t and u are both zero. As we have seen, the existence of such a point implies the existence of a complex root of the polynomial X. If this root is real, X has a linear factor, and the process can be continued. If the root is not real, X has a quadratic factor, and the process can also be continued.

Suppose no point with $t = u = 0$ exists, then $t^2 + u^2$ is always different from zero, and the function

(43) $$y = \frac{(t^2 + u^2)(t\, t'' + u\, u'') + (t\, u' - u\, t')^2 - (t\, t' + u\, u')^2}{r(t^2 + u^2)^2}$$

is everywhere finite. Note that the factor r in the denominator cancels out, because t', u', t'', u'' are divisible by r. Now Gauss considers the double integral

(44) $$\Omega = \iint y\, dr\, d\varphi,$$

integrated from $r = 0$ to $r = R$ and from $\varphi = 0$ to $\varphi = 360°$. One can integrate first with respect to r or first with respect to φ: the result is the same. The indefinite integral with respect to φ is

(45) $$\int y\, d\varphi = \frac{t\, u' - u\, t'}{r(t^2 + u^2)}$$

for if one differentiates the function on the right with respect to φ, one obtains just y. Now the function on the right has the same value for $\varphi = 0$ as for $\varphi = 360°$, so the integral on the left, taken from 0 to 360, is zero. This implies

(46) $$\Omega = 0.$$

On the other hand, if one integrates first with respect to r, one obtains the indefinite integral

(47)
$$\int y\,dr = \frac{t\,t' + u\,u'}{t^2 + u^2}.$$

For $r=0$ this expression is zero, and for $r=R$ it is positive, as we have seen. So the integral on the left, extended from 0 to R, is positive, and hence Ω is positive, contrary to (46). So the hypothesis that t and u are never both zero leads to a contradiction.

It is very easy to follow the proof of Gauss step by step. But how did he find his proof? In particular, how did he find the complicated expression y defined by (43)? I don't know, but I can make a guess.

We may consider $X = t + iu$ as a function of the complex variable

(48)
$$z = r(\cos \varphi + i \sin \varphi).$$

Geometrically speaking, the function

$$X(z) = X(r, \varphi) = t + iu$$

defines a mapping of the z-plane into the X-plane. In the X-plane we may also introduce polar coordinates:

(49)
$$X = s(\cos \beta + i \sin \beta).$$

We now have

(50)
$$\operatorname{tg} \beta = \frac{u}{t}.$$

Differentiating (50) with respect to φ and r one obtains

(51)
$$\frac{\partial \beta}{\partial \varphi} = \cos^2 \beta \frac{t\,t' + u\,u'}{t^2} = \frac{t\,t' + u\,u'}{t^2 + u^2} = U$$

and

(52)
$$\frac{\partial \beta}{\partial r} = \cos^2 \beta \frac{t\,u' - u\,t'}{r\,t^2} = \frac{t\,u' - u\,t'}{r(t^2 + u^2)} = V.$$

Differentiating once more, one obtains

(53)
$$\frac{\partial U}{\partial r} = \frac{\partial V}{\partial \varphi} = y.$$

Gauss himself makes use of the equations (53), from which he concludes

$$\int y\,dr = U \quad \text{and} \quad \int y\,d\varphi = V.$$

So it seems quite possible that he arrived at his complicated function y by differentiating U with respect to r and V with respect to φ. He could know beforehand that $\partial U/\partial r$ and $\partial V/\partial \varphi$ are equal, because U and V are the derivatives of one and the same function β with respect to φ and r.

It is true that the angle β is not uniquely defined: it is defined only modulo 2π, but in a neighbourhood of any point (r, φ) the angle β is a differentiable function of r and φ, so the total differential

$$d\beta = U\, d\varphi + V\, dr$$

is well-defined. It is possible that Gauss wanted to avoid the use of multivalued functions like β, and that this was the reason why he used only the derivatives U, V and y in his proof.

We now ask: How does the angle β vary when the point z moves on a large circle $r = R$ in the z-plane?

In his first proof, Gauss has shown: if φ goes from 0 to 2π, and if R is sufficiently large, the point X passes m times through the first, second, third, and fourth quadrants (in this order), which implies that β goes from zero to $m \cdot 2\pi$. Now let us calculate $\int d\beta$. Because of (51), we obtain

(54) $$\beta(2\pi) - \beta(0) = \int d\beta = \int U\, d\varphi.$$

This difference must be a multiple of 2π. Since U is positive, it is a positive multiple, say $m' \cdot 2\pi$. From Gauss' first proof we know that m' is equal to m, the degree of the polynomial X, but Gauss does not need this result. For his present proof, it is sufficient to know that $\int U\, d\varphi$ is positive.

As long as $t^2 + u^2$ is not zero, the integral (54) is a continuous function of r, so the integer m' cannot change, and the integral $\int U\, d\varphi$ remains constant. On the other hand, it is positive for $r = R$ and zero for $r = 0$. Thus one obtains a contradiction.

It is quite possible that Gauss had this simple proof in mind. However, he found a way to avoid the use of the multivalued angle β, writing the integral $\int U\, d\varphi$ as a double integral

$$\int U\, d\varphi = \iint y\, dr\, d\varphi$$

and interchanging the order of the integrations. Because $\int y\, d\varphi$ is always zero, the double integral is zero, but on the other hand $\int U\, d\varphi$ is positive for sufficiently large R. Thus Gauss obtained a contradiction.

Chapter 6
Evariste Galois

Evariste Galois was born in October 1811. Twenty years and seven months later he died in a duel. In the meantime he had created one of the most important and beautiful theories in the history of algebra: the Theory of Galois.

The dramatic story of his life is well known. One may consult the classical biography of Paul Dupuy: "La vie d'Evariste Galois" in Annales de l'Ecole Normale, série 3, Vol. 13, p. 197–266, or the more recent biography "Evariste Galois" by L. Kollros, also in French, in "Kurze Mathematiker-Biographien", Birkhäuser-Verlag, Basel 1978.

The Work of Galois

The mathematical works of Galois have been published first in 1846 by Liouville in his Journal de Mathématiques. They were reprinted in 1897 by Gauthier-Villars with an introduction by Emile Picard. A more complete, critical edition of all preserved letters and manuscripts, prepared by R. Bourgne and J.-P. Azra, was published by Gauthier-Villars in 1962 under the title "Ecrits et mémoires mathématiques d'Evariste Galois".

Galois' first published paper was an article of eight pages on continued fractions in Annales de Mathématiques de Gergonne, Vol. 19, p. 294–302 (1828). In this paper, Galois proved:

If one of the roots of an equation of arbitrary degree (with rational coefficients) is an immediately periodic continued fraction, then another root is also a periodic continued fraction, which one obtains by dividing −1 by the same continued fraction, written in the inverse order.

This is a nice addition to the results of Euler and Lagrange on continued fractions.

In May 1829, Galois presented a first account of his investigations on the solution of algebraic equations to the Académie des Sciences de Paris. A second memoir, on equations of prime degree, was presented eight days later, on June 1. Both papers were sent to Cauchy, who lost them. They have never been found.

In February 1830, Galois presented to the Academy another memoir on the solution of algebraic equations. This time the Academy gave the paper to its perpetual secretary Fourier. But Fourier died before he could examine the paper. The manuscript has not been found among his papers.

In April 1830 a short note of Galois was published in the Bulletin des Sciences mathématiques of Férussac (Oeuvres mathématiques de Galois, 1897, p. 11-12), in which some of the main results of his Academy memoir were announced without proofs. The first and most important theorem announced in this paper reads:

In order that an equation of prime degree be solvable by radicals, it is necessary and sufficient that, if two of its roots are known, the others can be expressed rationally.

This theorem implies that the general equation of degree 5 cannot be solved by radicals.

In the same year 1830, Galois published two more little papers on questions of analysis and on the numerical solution of equations (Oeuvres, 1897, p. 9-10 and 13-14).

Of great importance for the history of modern algebra is another paper "Sur la théorie des nombres", published in Férussac's Bulletin in June 1830 (Oeuvres, 1897, p. 15-23), in which Galois determined the structure of *finite fields*. An account of the contents of this paper will be given at the end of the present chapter.

In January 1831 the Academy received a third, revised version of Galois' great memoir. It was entitled "Mémoire sur les conditions de résolubilité des équations par radicaux". The text of this memoir can be found in the Oeuvres mathématiques (1897), p. 33-50. A critical edition with marginal notes, including corrections made by Galois himself, was published in "Ecrits and mémoires mathématiques d'Evariste Galois" (Paris 1962), p. 37-109.

The Academy asked the members Poisson and Lacroix to write a report on the manuscript. Poisson examined it carefully, but he declared that he could not understand it. The complete text of Poisson's report has been published by René Taton in a paper entitled "Les relations d'Evariste Galois avec les mathématiciens de son temps", Revue d'histoire des sciences et de leurs applications I (1947), p. 114-130. The report ends thus (my translation):

We have done our utmost to understand the demonstrations of Galois. His reasonings are not sufficiently clear, nor are they developed so far that we could judge their exactness, and we are not even able to give an idea of his reasoning in the report. The author states that the proposition which is the special object of this memoir is a part of a general theory, which is susceptible of many applications. It often happens that several parts of a theory, elucidating each other, are easier to grasp as a whole than isolated. Therefore, to form a definite opinion, one might wait until the author will have published his work as a whole. But in the present state of the part submitted to the Academy we cannot propose to give it your approbation.

The Duel

Galois had no opportunity to explain his complete theory. During these years he took part in numerous anti-monarchist riots. He was imprisoned twice. In May 1832 he was forced to accept a duel. He was sure that he would be killed. In a letter to his political friends he wrote:

"Je meurs victime d'une infâme coquette."

It has been said that his opponent wanted to kill him in order to get rid of a dangerous republican, but this rumour is not confirmed by Galois' own testimony, for at the end of his letter he writes.

"Pardon pour ceux qui m'ont tué, ils sont de bonne foi."

I suppose Galois himself had more information about the motives of those who challenged and killed him than his friends, who suspected political motives. When he says "ils sont de bonne foi" he must have been convinced that their motives were sincere. In a letter written to two of his friends on the eve of the duel he writes

Votre tâche est bien simple: prouver que je me suis battu malgré moi, c'est-à-dire après avoir épuisé tout moyen d'accommodement, et dire si je suis capable de mentir, de mentir même pour un si petit objet que celui dont il s'agissait.

and in another letter to his political friends

Je prends le ciel à témoin que c'est contraint et forcé que j'ai cédé à une provocation que j'ai conjurée par tous les moyens.

Je me repens d'avoir dit une vérité funeste à des hommes si peu en état de l'entendre de sang-froid. Mais enfin j'ai dit la vérité.

J'emporte au tombeau une conscience nette de mensonge, nette de sang patriote.

The duel took place on May 30 1832. He used the night before the duel to write a long letter to his friend Auguste Chevalier, in which he explained the fundamental ideas of his theory. This letter was published in September 1832 in the Revue encyclopédique, and republished in the Oeuvres (1897), p. 25–32. It ends thus:

Tu feras imprimer cette Lettre dans la Revue encyclopédique.

Je me suis souvent hasardé dans ma vie à avancer des propositions dont je n'étais pas sûr; mais tous ce que j'ai écrit là est depuis bientôt un an dans ma tête, et il est trop de mon intérêt de ne pas me tromper pour qu'on me soupçonne d'énoncer des théorèmes dont je n'aurais pas la démonstration complète.

Tu prieras publiquement Jacobi et Gauss de donner leur avis, non sur la vérité, mais sur l'importance des théorèmes.

Après cela, il y aura, j'espère, des gens qui trouveront leur profit à déchiffrer tout ce gâchis.

Je t'embrasse avec effusion.

Le 29 mai 1832.   E. Galois.

The next morning Galois was shot.

The Memoir of 1831

For us, who have learnt Galois Theory from a textbook or from a course of lectures, it is not so difficult to understand Galois' memoir as it was for Poisson.

Galois starts with an equation $f(x) = 0$. The coefficients are supposed to be known quantities, for instance rational or irrational numbers or just letters. All rational functions of these coefficients are called *rational*. One may also *adjoin* other quantities, for instance m-th roots of rational quantities, and consider as rational in a larger sense all rational functions of these quantities, says Galois. In modern terminology one would say that a certain "ground field" is presupposed, which may be extended by adjunctions in the course of the investigations.

If a polynomial $f(x)$ can be factored without leaving the ground field, it is called *reducible*, otherwise *irreducible*.

As a rule, but not consistently, Galois uses the words *permutation* and *substitition* in the same sense as Cauchy. A permutation is an ordering of a finite set, and a substitution is a passage from one ordering to another (or the same) ordering.

Galois now considers *groups* of substitutions having the property: if S and T belong to the group, so does ST.

If a polynomial f has a root in common with an irreducible polynomial g, then f is divisible by g. This is Galois' first lemma. It is also the first theorem in the 1829 memoir of Abel. The lemma implies that the field extension $K(V)$ obtained by adjoining a root V of an irreducible polynomial $g(x)$ is completely known as soon as the ground field K and the polynomial g are known. In modern terminology the field $K(V)$ is isomorphic to the residue class ring $K[x]/(g)$.

Galois next proves: If an equation $g(x)=0$ has no multiple roots and if a, b, c, ... are its roots, one can always form a function V of the roots such that all values of V obtained by permuting the roots are different.

For instance, one can take

(1) $$V = Aa + Bb + Cc + ...$$

with conveniently chosen integers A, B, C, ..., says Galois.

From this lemma, Galois deduces a special case of what we now call the "Theorem of the Primitive Element":

Lemma 3. If V is chosen as before, all roots a, b, c, ... are expressible as rational functions of V.

To prove this important lemma, Galois puts

$$V = \varphi(a, b, c, ...).$$

He now permutes the roots b, c, ... in all possible ways, keeping fixed only root a, and forms the product

$$[V - \varphi(a, b, c, ...)] \cdot [V - \varphi(a, c, b, ...)] \cdot ....$$

This is a symmetric function of b, c, ..., which are the roots of the polynomial

$$g(x)/(x-a),$$

hence it can be expressed as a rational function of a. So we have an equation

(2) $$F(V, a) = 0.$$

This equation and

(3) $$g(a) = 0$$

have in common only one root a, for it cannot happen that, for instance, $F(V, b)$ is zero, says Galois.

Now if two equations like (2) and (3) have only one root a in common, this root can be computed rationally. Hence a is a rational function of V.

Galois is right in saying that $F(V, b)$ cannot be zero, for $F(V, b)$ is a product of factors

$$[V - \varphi(b, a, c, \ldots)] \cdot [V - \varphi(b, c, a, \ldots)] \cdot \ldots$$

in which the permutations $(b, a, c, \ldots)$ etc. are all permutations of $(a, b, c, \ldots)$ in which b comes first, while the others $(a, c, \ldots)$ are permuted in all possible ways. This follows from the definition of $F(V, a)$, as H.M. Edwards has pointed out in his book "Galois Theory" (Springer-Verlag 1984), p. 44–45. Namely: since all expressions $\varphi(b, a, c, \ldots)$ etc. are supposed to be different from $V = \varphi(a, b, \ldots)$, it follows that $F(V, b)$ is different from zero, and so are $F(V, c)$, etc.

Poisson made a marginal note to Lemma 3, saying: "The proof of this lemma is insufficient, but it is true by article 100 of the memoir of Lagrange." It is easy to understand Poisson's attitude. Galois' proof is only a sketch, and did not elaborate his statement that $F(V, b)$ is not zero. Poisson's last statement "It is true by article 100 of Lagrange" is correct, for in article 100 of Lagrange's "Réflexions" a complete proof of the lemma is given.

In my opinion Galois was right in claiming that his proof is essentially correct, but Poisson was right in declaring that it is incomplete.

In modern notation we may now write

$$(4) \qquad\qquad K(a, b, c, \ldots) = K(V)$$

where K is the ground field. The "primitive element" V is a root of an irreducible equation. Let

$$V, V', V'', \ldots, V^{(n-1)}$$

be the roots of this equation. Lemma 4 says: If $a = \varphi(V)$ is a root of the original equation, $\varphi(V')$ will also be a root. The proof is easy.

Next comes the main theorem:

Proposition I. There is a group of permutations of the letters a, b, c, ... such that

1° Every function of the roots, invariable under the substitutions of the group, is rationally known;

2° conversely, every function of the roots rationally known is invariable under the group.

Galois' terminology is not consistent. He first speaks of "permutations" and next of "substitutions" forming the group, but what he wants to say is completely clear.

To prove this theorem, Galois expresses the roots as rational functions of V:

$$\varphi V, \varphi_1 V, \ldots, \varphi_{m-1} V.$$

He next writes down the permutations

$$
\begin{array}{llll}
\varphi V, & \varphi_1 V, & \varphi_2 V, & \ldots, \varphi_{m-1} V \\
\varphi V', & \varphi_1 V', & \varphi_2 V', & \ldots, \varphi_{m-1} V' \\
\ldots & \ldots & \ldots & \ldots \\
\varphi V^{(n-1)}, & \varphi_1 V^{(n-1)}, & \varphi_2 V^{(n-1)}, & \ldots, \varphi_{m-1} V^{(n-1)}
\end{array}
$$

and he states that the "group of permutations" (meaning the corresponding group of substitutions) satisfies the required conditions. The proof is very short, but it is not difficult for a modern reader to elaborate the single steps.

Galois next investigates how the group of the equation changes when the ground field is extended by the adjunction of a root or of all roots of an auxiliary equation. It is clear that after the adjunction the Galois group will be a subgroup H of the original group G. If H is a proper subgroup, G can be decomposed as follows:

$$
(5) \qquad\qquad G = H + HS + HS' + \ldots
$$

or, alternately, as

$$
(6) \qquad\qquad G = H + TH + T'H + \ldots.
$$

These two decompositions are most clearly explained in the letter to Chevalier (Oeuvres de Galois, 1897, p. 25–32).

The two decompositions do not always coincide, says Galois. If they do coincide, the decomposition is called "proper". In modern terminology, this is the case when H is an "invariant subgroup", or "normal divisor" of G. In particular, if *all* roots of an auxiliary equation are adjoined, the two decompositions will coincide. This is Proposition III of Galois. The proof is omitted ("On trouvera la démonstration").

Galois now comes to his main problem: In what case is an equation solvable by radicals?

One can, of course, restrict oneself to radicals of prime degree p. Every time a p-th root is extracted, Galois supposes the p-th roots of unity to be adjoined beforehand. This is not an essential restriction, because Gauss had proved already that the p-th roots of unity can be expressed by means of radicals of degrees less than p.

Let us suppose that the adjunction of a radical r, root of an equation

$$
(7) \qquad\qquad x^p - s = 0,
$$

leads to a reduction of the Galois group. Because the p-th roots of unity

$$
\alpha, \alpha^2, \ldots, \alpha^p = 1
$$

are in the ground field, the same reduction is obtained by adjoining *all* roots of the equation (7). By Proposition III, the decomposition (5) will be a proper decomposition, that is, the subgroup H is a normal divisor. Galois stated, but did not prove, that the number of terms in the decomposition (5) (which we call the *index* of H in G) is just the prime number p. Conversely, if G has a normal divisor H of prime index p, one can reduce the Galois group G to the subgroup H by adjoining a radical of degree p. This is proved as in our textbooks by taking a function θ invariant under the subgroup H and forming a "Lagrange resolvent"

(8)
$$z = \theta + \alpha\theta_1 + \alpha^2\theta_2 + \ldots + \alpha^{p-1}\theta_{p-1}$$

where α is a p-th root of unity, while θ_1, θ_2, ... are obtained from θ by the substitutions

$$S, S^2, \ldots, S^{p-1}$$

representing the cosets in the decomposition (5).

It follows that an equation $g(x) = 0$ is solvable by radicals if and only if a sequence of subgroups

$$G \supset H_1 \supset H_2 \supset \ldots \supset H_m = E$$

exists, such that every H_k is a normal divisor of the preceding H_{k-1} or G, while all indices are prime. If this is the case we say that the group G is *solvable*.

Galois next supposes that the equation $f(x) = 0$ is irreducible and of prime degree n. He proves: The equation can be solved by radicals if and only if each of the substitutions of G transforms x_k into $x_{k'}$ by a linear transformation of k modulo n:

$$k' \equiv ak + b \pmod{n}.$$

The Galois group of the general quintic equation is not of this form, hence this equation cannot be solved by radicals. Thus, Abel's result follows from the theory of Galois.

In the last version of his Academy memoir Galois quoted Abel, but at the time when he sent his first version to the Academy he did not even know the name of Abel. His main sources were the works of Lagrange, Gauss, and Cauchy.

Galois Fields

Both Abel and Galois had a clear notion of what we now call "field". Galois states right at the beginning of his great memoir:

"One can agree to consider as rational every rational function of a certain number of quantities regarded as known a priori", and he goes on to explain what he means by adjoining a certain quantity to the field of quantities considered as known.

The fields considered by Abel and Galois in their papers on the resolution of equations all contain the rational number field. In modern terminology they are fields of characteristic zero. If the characteristic were p, the equation

$$x^p - 1 = 0$$

would have only one root $x = 1$, whereas Abel and Galois always suppose that the p-th roots of unity are all different.

However, in his paper "Sur la théorie des nombres", which was published in 1830 in the Bulletin des Sciences de Férussac (Oeuvres de Galois, Paris 1897, p. 15–23) Galois constructs finite fields, the so-called *Galois-Fields*. He states from the very beginning that his object is to consider algebraic structures in which all quantities, multiplied by p, are considered to be zero. In his own words, translated into English, he says:

> If one agrees to regard as zero all quantities which, in algebraic calculations, are found to be multiplied by p, and if one tries to find, under this convention, the solution of an algebraic equation $Fx = 0$, which Mr. Gauss designates by the notation $Fx \equiv 0$, the custom is to consider integer solutions only. Having been led, by my own research, to consider incommensurable solutions, I have attained some results which I consider new.

From these words it is clear that the starting point of Galois was the calculus of congruences modulo a prime p, initiated by Gauss. It was known that residue classes modulo p can be added, subtracted, and multiplied, and that the congruence

$$ax \equiv b \quad (\text{modulo } p)$$

can always be solved in rational integers, provided a is not congruent to zero. In other words, the residue classes modulo p form a field.

Gauss had also considered congruences of higher degrees such as

$$x^2 \equiv a \quad (\text{modulo } p),$$

but he admitted rational solutions only. Galois now asks whether one can introduce irrational solutions, that is, whether one can enlarge the residue class field by the adjunction of roots not contained in the original field.

Galois supposes the polynomial Fx to be irreducible modulo p. He asks whether one can solve the congruence $Fx \equiv 0$ by introducing new "symbols", which might be just as useful as the imaginary unit i in ordinary analysis.

Galois calls i one of the roots of the congruence $Fx \equiv 0$ of degree v. He forms the p^v expressions

(A) $a + a_1 i + a_2 i^2 + \ldots + a_{v-1} i^{v-1},$

where $a, a_1, a_2, \ldots, a_{v-1}$ are integers modulo p. These p^v elements form what we call today a "Galois Field" $GF(p^v)$.

It is easy to show that the expressions (A) form a field, that is, that they satisfy the well known rules of addition, subtraction, multiplication, and division.

Galois now takes an element α of the form (A), in which the coefficients a, $a_1, \ldots, a_{v-1}$ are not all zero. The powers α, α^2, ... cannot be all different, hence a power α^n must be equal to 1. If n is the smallest integer for which α^n is 1, the expressions

$$1, \alpha, \alpha^2, \ldots, \alpha^{n-1}$$

must be all different. In modern terminology, they form a subgroup of the multiplicative group of the Galois field.

Multiplying these numbers by another element $\beta \neq 0$, one obtains a coset of the subgroup. Going on in the same way, one finds that all cosets together form the whole multiplicative subgroup of order $p^v - 1$, and that the exponent n is a divisor of $p^v - 1$. Hence one has

$$\alpha^{p^v - 1} = 1.$$

Next one proves, says Galois, as in the theory of residue classes modulo p, that there exist "primitive roots" for which n is exactly $p^v - 1$. All other non-zero elements of the Galois field are powers of a primitive element α. The proof of the existence of such an element, given by Gauss for the case of the residue class field modulo p, works just as well in the case of $GF(p^v)$.

We now see that all elements of the Galois field, including zero, are roots of the polynomial

(B) $$x^{p^v} - x$$

and that every irreducible polynomial Fx of degree v is a divisor of the polynomial (B). If α is one of the roots of such a polynomial, the others are

$$\alpha^p, \alpha^{p^2}, \ldots, \alpha^{p^{v-1}}.$$

This follows from the well known congruence

$$(Fx)^p \equiv F(x^p).$$

At the end of his treatise, Galois reverses the situation. He starts with any field extension of $GF(p)$ in which the polynomial (B) can be completely factorized. Restricting himself to the subfield generated by the roots, he takes a "primitive element" i of the subfield. Such an element always exists according to a theorem known to Abel, says Galois. Every such i is a root of a (mod p) irreducible polynomial Fx. No matter which irreducible polynomial of degree v one chooses, one always obtains the same field $GF(p^v)$. In most cases, the simplest way to obtain such a polynomial is "par tatonnement", says Galois, by trial and error. As an example, he takes $p=7$ and $v=3$. The polynomial $x^3 - 2$ is irreducible (mod 7), and a root i of this polynomial generates the field $GF(7^3)$.

The Publication of Galois' Papers

Galois' "Mémoire sur les conditions de résolubilité des équations par radicaux" was published in 1846, fourteen years after the death of Galois, by Liouville in his Journal de mathématiques pures et appliquées 11, p. 381–444. In an "Avertissement" preceding the memoir, Liouville reproduces the letter of Galois to Chevalier and adds:

Inserting in their Recueil the letter one has just read, the editors of the *Revue encyclopédique* announced that they would soon publish the manuscripts left by Galois. But this promise has not been kept. However, Monsieur Auguste Chevalier has prepared the work. He has given us, and one will find in the pages to follow:

1° A memoir on the conditions of solvability of equations by radicals, with an application to equations of prime degree,

2° A fragment of a second memoir, in which Galois treats the general theory of those equations which he calls *primitive*.

We have preserved most of the notes which Monsieur Auguste Chevalier had added to the memoirs just mentioned. These notes are all marked A.Ch. The unsigned notes are by Galois himself.

We will complete this publication by some other fragments drawn from the notes of Galois. Without having great importance these notes might still arouse the interest of the geometers.

The notes mentioned by Liouville in the last sentence have been published by Jules Tannery (Manuscrits de Evariste Galois, Gauthier-Villars, Paris 1908), and again in "Ecrits et mémoires d'Evariste Galois" (Paris 1962).

Liouville also states that he experienced great joy when he realized, after filling a few slight gaps, that the method by which Galois proved his beautiful theorem was completely accurate.

For more information on the activities of Liouville, Hermite, and Serret during the years 1846–1854 I may refer to a very interesting paper by B.M. Kiernan in the Archive for History of Exact Sciences 8, p. 40–154 (1971) entitled "Galois Theory from Lagrange to Artin". Section 11 of this paper deals with the publication of Galois' papers and the reaction of the French mathematicians to this publication. The next section of the present chapter is mainly drawn from Kiernan's paper.

Hermite, Puiseux, and Serret

The French mathematician Charles Hermite (1822–1882) was a pupil of the same Louis Richard who taught Galois. In 1842, at the age of twenty, he published a paper "Considérations sur la résolution algébrique de l'équation du cinquième degré", in which he sketched with great clarity and precision Lagrange's ideas concerning the general quintic equation.

In 1847 or earlier Hermite wrote a letter to Jacobi, in which he mentioned Galois' work on elliptic functions. So Hermite was acquainted with the work of Galois at a date shortly after the publication of the memoirs of Galois in 1846 or even earlier.

In 1850, Victor Puiseux published a fundamental paper entitled "Recherches sur les fonctions algébriques" in Vol. 15 of Liouville's "Journal de mathématiques pures et appliquées", p. 365–480. In this paper, Puiseux considers an algebraic function w of a complex variable z defined by an equation

$$(9) \qquad\qquad\qquad f(z, w) = 0$$

in which $f(z, w)$ is a polynomial in w, irreducible in the field of rational functions of z. In the neighbourhood of any point z_0 which is not a "branch point", the roots w_1, w_2, ..., w_n of the equation (9) can be expanded as convergent power series in $z - z_0$. If z_0 is a branch point, one has to use powers of $z - z_0$ with fractional exponents. These power series are called up to the present day "Puiseux series".

If one starts with a non-branch point z_0 and if one makes z move on a closed path, avoiding the branch points and ending at z_0, one will end up with a permutation of the original roots w_1, ..., w_n. These permutations obviously form a group, but Puiseux does not use the word "groupe". In Jordan's "Traité des substitutions" this group is called "groupe de monodromie".

In the following year 1851, Hermite published a short paper in which he showed that the group of substitutions of the roots w_1, ..., w_n considered by Puiseux is just the Galois group of the equation (9), if $\mathbb{C}(z)$ is the ground field. Thus, an important link between Galois theory and complex function theory was established (Oeuvres d'Hermite I, p. 276–280).

Liouville conducted a series of seminars on Galois theory. The seminar was attended by Joseph-Alfred Serret, the author of a very influential "Cours d'algèbre supérieure". In the first edition of this textbook (published in 1849) one finds a proof of the fundamental third lemma of Galois, which reads in the text of Serret (3rd edition, Vol. 2, p. 413):

If

$$f(x) = 0$$

is an equation of degree n, which has no equal roots, and if

$$V = \varphi(x_0, x_1, \ldots, x_{n-1})$$

is a rational function of the roots x_0, x_1, ..., x_{n-1}, chosen in such a way that the $1 \cdot 2 \cdot \ldots \cdot n$ values which it takes when the roots are permuted are all different, then one can express the roots x_0, x_1, ..., x_{n-1} as rational functions of V.

The proof given by Serret is just the proof of Galois, which we have discussed earlier.

Some time before 1854 Hermite sent to Serret a proof of a theorem of Galois which says: An irreducible equation of prime degree is solvable by radicals only if all roots can be rationally expressed by any two of the roots. Serret inserted this proof in the second edition of his "Cours d'algèbre supérieure", which came out in 1854. Serret also included a French translation of a paper of Kronecker "Über die algebraisch auflösbaren Gleichungen", presented to the Berlin Academy in 1853 (Kronecker's Werke IV, p. 1–11).

The third edition of Serret's "Cours d'algèbre" (1866) contains a thorough exposition of the theory of Galois. Serret says

"I have followed the order of the propositions which Galois adopted, but very often I had to fill out the inadequate proofs."

According to Kiernan, Serret's main contribution is notational. For Serret, the Galois group is a group of *substitutions* in the sense of Cauchy. He introduces the notation of what we today call "conjugate subgroups". If H is a subgroup consisting of m substitutions

$$1, S_1, S_2, \ldots, S_{m-1},$$

then a conjugate subgroup consists of

$$1, TS_1T^{-1}, TS_2T^{-1}, \ldots, TS_{m-1}T^{-1}.$$

Serret's textbook was very influential. Many editions followed the third edition. A German translation by G. Wertheim was published in 1868.

We now return to the year 1852, and turn to Italy.

Enrico Betti

The first to present an exposition of Galois theory according to the ideas of Galois, but with more complete proofs, was Enrico Betti, a very interesting personality in the history of algebra and algebraic topology. His name is known to topologists because it is connected with the so-called "Betti-numbers" or "homology numbers".

Actually, these numbers were not invented by Betti, but by Henri Poincaré, who was inspired by a paper of Betti. The history of this invention has been studied in a paper by Maja Bollinger "Entwicklung des Homologiebegriffs" in Archive for History of Exact Sciences 9, p. 94–170 (1972). It is worthwile to summarize the main facts about "Betti numbers" here.

In Riemann's "Gesammelte mathematische Werke" (1871) one finds a posthumous "Fragment on Analysis Situs", in which Riemann defines what we now call "homology numbers modulo 2", that is, numbers of linearly independent homology classes for non-oriented cycles with integers modulo 2 as coefficients. Riemann proves that these numbers are independent of the choice of the basic cycles. His method of proof is just the same by which Steinitz later proved that the degree of a finite field extension does not depend on the choice of the basis. One finds the proof in every textbook of modern algebra.

Riemann visited Betti in Italy. In a letter dated 21 January, 1871, he calls Betti "carissimo amico". It appears that Riemann explained his ideas concerning homology numbers to Betti.

In 1871 Betti published a paper "Sopra gli spazi di un numero qualunque di dimensioni" in Annali di matematica pura e applicata (2) 4, p. 140–158 (1871). The ideas underlying this paper are just the same as in the fragment of Riemann, and Betti says expressly that his proof of the independence of the homology numbers from the choice of the basis is the same as a proof given by Riemann for the special case of closed paths on a Riemann surface. However, Betti's proof is not correct, as Heegard and Maja Bollinger have shown. Betti misunderstood Riemann's ideas in several respects.

Poincaré, inspired by Betti's paper, developed a completely correct homology theory. He considered oriented cycles, multiplied by rational numbers, so his homology is a "homology with admitted division", as we say today. The numbers of linearly independent cycles of all dimensions less than m on a manifold of dimension m are called, in honour of Betti, "nombres de Betti". On an orientable closed surface, for instance on the Riemann surface of an algebraic function, these numbers happen to be equal to the homology numbers modulo 2 studied by Riemann and Betti, but in general they may be different.

Just as Betti elaborated Riemann's ideas on homology, he also elaborated the theory of Galois in a paper "Sulla risoluzione delle equazioni algebriche", published in 1852 (Opere matematiche I, p. 31–80). In the introduction to this paper he writes, referring to the work of Galois and Abel:

"The conditions for solvability by radicals of equations of prime degree can thus be held determined and proved by the procedures of both of these mathematicians. The conditions remain to be determined for equations of non-prime degree, and much on this is found proposed by Galois and Abel in different ways, but without proof, in their posthumous papers. To fill in these gaps is the main intention of my work" (Kiernan's translation).

In fact, Betti provided proofs for several theorems which Galois merely stated. A good example is given on page 107 of Kiernan's paper in Archive for History of Exact Sciences 8.

Betti is quite near to the modern notion of a quotient group. If H is a normal subgroup of G, Betti assumes that the representatives S_i of the cosets S_iH can be chosen in such a way that they form a group. In this particular case the group formed by the representatives S_i is isomorphic to what we call the quotient group G/H.

For more details on Betti's memoir I may refer to the paper of Kiernan.

Betti's exposition of Galois theory, being written in Italian, was not as influencial as that of Serret, whose "Cours d'algèbre supérieure" remained a standard text for a long time: the sixth edition appeared as late as 1928.

The Second Posthumous Memoir of Galois

As we have seen, Galois completely determined the structure of the Galois groups of solvable, irreducible equations of prime degree p. Since an equation is irreducible only if its group is transitive, and solvable only if the group is solvable, the theorem proved by Galois is equivalent to the following:

A transitive group of permutations of degree p (that is, on p letters) *is solvable only if its permutations can be written as*

$$k' \equiv ak + b \quad (\text{mod } p).$$

In a second, unfinished memoir published in 1846 (Oeuvres, p. 51–61) Galois investigates the more general case in which the group is supposed to be primitive.

A transitive group is called *primitive*, if it is impossible to divide the letters into several subsets of more than one elements each:

$$\{a_1, \ldots, a_m\}, \{a_{m+1}, \ldots, a_{2m}\}, \ldots$$

in such a way that the group transforms each subset either into itself or into another one of the subsets.

Obviously, a transitive group of prime degree p is always primitive. Galois now considers non-prime degrees, and he proves:

If a solvable group is primitive, its degree must be a power of a prime.

Galois' proof of this theorem is not easy to understand, but it is correct. Camille Jordan presented an elaborate proof in his "Traité des substitutions" (1870). In this monumental volume, Jordan developed an exhaustive classification of solvable primitive groups of degree p^n.

The theorem just mentioned had been announced already, together with other results, in Galois' published paper of 1830 entitled "Analyse d'un mémoire sur la résolution algébrique des équations" (Oeuvres, 1897, p. 11–12). His preliminary results on primitive groups were far surpassed by those of Jordan.

The remainder of the second memoir of Galois deals with primitive groups of degree p^2. The memoir is unfinished. It ends with a question and with the words "C'est ce que je vais rechercher".

Chapter 7
Camille Jordan

On the life and work of Camille Jordan (1838–1922) one may consult the excellent obituary of Henry Lebesgue, read at the Paris Academy in June 1923, and reprinted in Jordan's Oeuvres IV, p. X–XXIX.

Jordan was born at Lyon. At the age of 17, he was admitted to the Ecole Polytechnique as the first in rank, with 19.8 points out of 20. In 1876 he became professor at the same Ecole.

The name of Jordan is well known to all mathematicians of my generation because of his excellent "Cours d'analyse", a considerably enlarged elaboration of his lectures at the Ecole Polytechnique. As far as I know, this is the earliest textbook in which the whole of classical analysis is presented as a unified, completely logical theory. For instance, Jordan was the first to present, in the second and later editions of his Cours d'Analyse, clear definitions of the notions "volume" and "multiple integral", and he specified conditions under which a multiple integral can be evaluated by successive integrations. For me, every single chapter of the Cours d'analyse is a pleasure to read.

Jordan's Traité

Jordan's monumental work of 667 pages "Traité des substitutions et des équations algébriques", published in 1870 by Gauthier-Villars, is a masterpiece of mathematical architecture. The beauty of the edifice erected by Jordan is admirable.

In the preface to his Traité Jordan gives due credit to his predecessors: first of course to Galois, who "invented the principles of Galois theory", and to Betti, who wrote "an important memoir, in which the complete sequence of theorems of Galois has been rigorously established for the first time". Next, Jordan mentions the contributions of Abel, Hermite, Kronecker, and Brioschi concerning the Galois groups of certain division problems of elliptic and Abelian functions, and the investigations of the geometers Hesse, Cayley, Clebsch, Kummer, Salmon, and Steiner, who studied a multitude of geometrical problems to which the methods of Galois can be applied. Finally, he acknowledges his indebtedness to the "Cours d'algèbre" of Serret, saying:

It is the careful study of this book that initiated us into Algebra and inspired in us the desire to contribute to its progress.

When Jordan says that he was initiated into Algebra by the book of Serret, this is certainly true, but it is not the whole truth. An explanation of Galois theory is found only in the third edition of Serret's "Cours d'algèbre", which appeared in 1864, but three years earlier Jordan had already quoted Galois in his Thèse de doctorat (Oeuvres de Camille Jordan I, 1961, p. 1–82). Right in the first chapter of this Thèse Jordan introduces the notion "système conjugé" according to Cauchy. Such a system consists of substitutions A, B, ... and their products such as

$$A^\alpha B^\beta C^\gamma B^{\beta'} \ldots.$$

This system is, of course, a *group* in the sense of Galois, and in fact, in Chapter III of his Thèse, Jordan writes:

Il est facile de voir, en effect, que la condition nécessaire et suffisante pour qu'une équation soit irréductible, est que le système conjugé qui lui correspond et que Galois nomme son *groupe*, soit transitif.

It follows that Jordan's initiation into Galois theory is not only due to Serret, but also to Galois himself. Jordan must have seen the fundamental memoir of Galois before 1861, for his Thèse was published in 1861 in the Journal de l'école polytechnique 12, p. 113–194.

In 1870, when the Traité was published, Jordan was already famous. Scholars from all parts of Europe came to Paris to see him. Sophus Lie came from Norway and Felix Klein from Germany. Klein and Lie were good friends at that time: they lived in Paris in adjacent rooms from April to June 1870, and this stay was of decisive importance for their later work on continuous and discrete groups of transformations. Klein writes about Lie and himself:

We lived room-to-room and sought scientific inspiration mainly in personal contact, especially with younger mathematicians. I was greatly impressed by Camille Jordan, whose Traité des substitutions had just been published. It appeared to us as a book with seven seals (Felix Klein, Gesammelte math. Abhandlungen I, p. 51).

Most of Jordan's early papers on Galois theory and on groups of substitutions were incorporated in his Traité. An excellent commentary to this part of Jordan's work, by Jean Dieudonné, can be found in Volume I of Jordan's Oeuvres (1961).

Following the chronological order, I shall first discuss Jordan's work on groups of Euclidean motions, and next present a summary of his Traité.

On Groups of Motions

In 1867, Jordan published a short note in the Comptes Rendus of the Paris Academy entitled "Sur les groupes de mouvements", in which he announces a complete determination of all possible groups of displacements of rigid bodies in Euclidean 3-space. In 1868–1869 the enumeration of these groups was presented in a two-part memoir entitled "Mémoire sur les groupes de mouvements", Annali di Matematica 2, p. 167–215 and 322–345.

The displacements considered by Jordan are helicoidal motions not only of a limited rigid body, but of the whole space. Particular cases are the rotations and translations. A *group* of motions is defined to be a set containing the product AB of any two elements of the set. Jordan tacitly assumes his groups to contain the inverses A^{-1} as well. Moreover, he restricts himself to topologically *closed* sets. This restriction is evident already in the CR-note, for here Jordan announces a proposition which he calls "la plus essentielle et la plus délicate à établir", namely:

"Let P and P' be two motions chosen at will. One can in general, and with some exceptions, obtain any motion by a convenient combination of P and P'."

It is clear that products composed from P and P' form, in any case, a denumerable set and not the whole group. But if one takes the closure of the set of products, one obtains, apart from special cases, the whole group.

Jordan's work on groups of motions was inspired by the "Etudes cristallographiques" of Bravais. In fact, Jordan states that several important special cases of his enumeration problem had been treated already by Bravais. He also says that his problem of determining all groups of motions can also be formulated thus:

"To determine all systems of molecules which can be superposed to themselves in several ways."

Jordan first determines all closed groups of translations and next all closed groups of rotations having one fixed point in common.

The determination of the translation groups is easy. There are four types of discrete groups of translations, generated by 3 or 2 or 1 or no linearly independent translations. There are three types of continuous groups of translations, of dimensions 3 and 2 and 1. Combining a continuous group of dimension 1 or 2 with a discrete group, one finds three mixed types. So there are just 10 types of topologically closed groups of translations.

The closed groups of rotations are not so easy to find. Discrete groups of rotations are, according to Jordan:

1) cyclic groups $C_1, C_2, C_3, \ldots,$
2) dihedral groups $D_4, D_6, D_8, \ldots,$
3) the tetrahedral group,
4) the octahedral group,
5) the icosahedral group.

Continuous groups of rotations are
6) the group of all rotations having a common fixed point O,
7) the group of rotations about a fixed axis.

Combining the last group with a rotation inverting the axis, one obtains

8) the group consisting of all rotations about an axis a and all "full turns" about axes b drawn through O and perpendicular to a.

Next Jordan undertakes the gigantic task to find all closed groups of motions by combining the rotations with translations. In modern terminology his method may be explained thus:

From all helicoidal motions of the group G one may take the rotational parts, thus obtaining a group of rotations R and a morphism

$$G \to R.$$

The kernel of this morphism is the group T of all translations in G. Hence

$$R \cong G/T.$$

So, in order to obtain the group G, one has to find all possible extensions of the normal subgroup T such that the factor group is isomorphic to R. This is, in fact, Jordan's method.

Jordan's enumeration is not complete. Leonhard Sohncke noted, in the Historical Introduction to his book "Entwicklung einer Theorie der Kristallstrukturen" (Leipzig 1879), that all crystallographic groups listed in § 20 of Chapter 6 of his book are missing in Jordan's list. Sohncke found this lacuna because he had investigated, by quite a different method, the regular point systems in the plane.

In spite of this deficiency, Jordan's pioneering work is admirable. Sohncke used Jordan's method to determine all three-dimensional crystallographic groups preserving the orientation, thus paving the way towards the complete enumeration of all crystallographic groups by A. Schoenflies and E.S. von Fedorow in the years 1889–1891. For the history of these investigations see J.J. Burckhardt: Zur Geschichte der Entdeckung der 230 Raumgruppen, Archive for History of Exact Sciences 4, p. 235–246 (1967), and also J.J. Burckhardt: Der Briefwechsel von E.S. von Fedorow und A. Schoenflies 1889–1908, same Archive 7, p. 91–141 (1971).

Jordan's memoir was not the earliest paper concerning Euclidean displacements and their composition. As early as 1758, Euler published a paper "Du mouvement des corps solides autour d'une axe variable" (Opera omnia, series secunda, Vol. 8, p. 154–193). Among other things, Euler proved that every displacement of a rigid body can be expressed as a product of an axial rotation and a translation. To describe rotations, he introduced the "Euler angles", which are still used by physicists.

Recently, Jeremy Gray has called attention to a nearly forgotten paper of Olinde Rodrigues: "Des lois geométriques qui régissent les déplacements d'un système solide dans l'espace, et de la variation des coordonnées provenant de ces déplacements considérés indépendamment des causes qui peuvent les produire", Journal de Math. (1) 5, p. 380–440 (1840). In this paper, Rodrigues proves that every displacement of a rigid body is the resultant of a rotation and a translation along the axis of the rotation. Jordan considers this result as well known, but he does not mention Rodrigues.

Following Euler, Rodrigues describes a rotation by four parameters g, h, l, θ, the first three determining the direction of the axis. He develops explicit formulae for the resultant of two rotations, and he stresses the fact that this product is not commutative. See J.J. Gray: Olinde Rodrigues' paper of 1840 on Transformation Groups, Archive for History of Exact Sciences 21, p. 375–385 (1980)

We now come to Jordan's Traité des substitutions.

On Congruences

The first "Book" of Jordan's Traité is entitled "Des congruences". In this book Jordan first summarizes the main results of Fermat and Gauss on congruences between integers and on power residues. Next he expounds, following Galois, the structure theory of what we call Galois Fields $GF(q)$.

Transitive and Primitive Groups of Substitutions

More than one-third of Jordan's Traité is occupied by Book 2: "On Substitutions". In Chapter 1 of this book, Jordan deals with substitutions (or, as we now call them, permutations) in general. He proves that the order of a subgroup H of a group G is a divisor of the order of the whole group. Following Cauchy, he proves that a group whose order is divisible by p contains an element of order p.

Next, Jordan defines the notions "transitive" and "primitive". A group is called "k-fold transitive" or "transitive of order k", if it transforms any k distinct letters into any other k distinct letters. Jordan mentions a 5-fold transitive group of substitutions on 12 letters discovered by Mathieu, and he proves some theorems on the order of transitivity. The work of Mathieu will be discussed in the next chapter together with the work of Cauchy, Serret, and Jordan on the possible values of the index i of a subgroup of S_n.

For a summary of later papers of Jordan on primitive and on multiply transitive subgroups of S_n see J. Dieudonné, Vol. I of Jordan's Oeuvres, pages XXX to XXXIII.

Series of Composition

A *composition series* of a group G is a sequence of groups

$$G \supset H \supset H' \supset \ldots \supset I$$

in which every term is a normal subgroup of the preceding one, and in which no intermediate normal subgroups can be inserted. Jordan proves that the quotients of the orders of successive groups are uniquely determined by the group, apart from their order. This fundamental theorem occurs already, though without proof, in Jordan's "Commentaire sur Galois" in Math. Annalen 1 (1869) on p. 152.

In 1889 *Otto Hölder* proved the stronger "Jordan-Hölder Theorem", which says that the factor groups

$$G/H, \; H/H', \ldots$$

are uniquely determined but for their order and but for isomorphisms.

Our modern definition of the notion "factor group G/H" by means of cosets is due to Hölder. However, the same notion occurs implicitly in an 1875 paper of Jordan entitled "Sur la limite de transitivité des groupes non alternés" (Oeuvres I, p. 365–396). On p. 371 of this paper Jordan defines:

> Two substitutions s and t permutable with a group H are called *congruent modulo H*, if one has
>
> $$s = th$$
>
> were h is a substitution in H.

Jordan now considers a sequence of substitutions $s_1, s_2, \ldots$ permutable with H and incongruent modulo H, such that for all α and β one has a relation of the form

$$s_\alpha s_\beta \equiv s_\gamma \quad \text{mod } H.$$

In this ease Jordan says that $s_1, s_2, \ldots$ form "un groupe suivant le module H", and he denotes this group by G/H, where G is the group generated by $s_1, s_2, \ldots$. Contrary to our usage, he does not require H to be contained in G. In our notation we would denote Jordan's quotient group by

$$G/(G \cap H),$$

and indeed, Jordan states that the order of G is the product of the orders of the factor group G/H and the intersection of G and H.

Jordan next defines "isomorphisms", which are what we today call homomorphisms or just morphisms. One-to-one morphisms are called by Jordan *holoedric*, the others *meriedric*.

Next Jordan considers the alternating group A_n. He shows that it is generated by cycles $(a\,b\,c)$, and that it is the only non-trivial normal subgroup of S_n, except for $n=4$. A corollary says: The group A_n is simple if n exceeds 4.

Linear Substitutions

The extremely interesting Chapter 3 of Book 2 of Jordan's Traité (p. 88–249) deals with what Jordan calls linear substitutions, and what we write in matrix form as

$$x' = A\,x.$$

In most cases, the field of coefficients is a prime field $GF(p)$, but in some cases, the prime field is extended to a Galois field $GF(p^\nu)$. For instance, on p. 114–126, in order to reduce the matrix A to the well-known "Jordan Normal Form", Jordan had to adjoin to the ground field the roots of the "characteristic equation"

$$\text{Det}\,(A - \lambda I) = 0.$$

On the history of the Jordan Normal Form see Thomas Hawkins: "Weierstrass and the Theory of Matrices", Archive for History of Exact Sciences 17, p. 119–163 (1977). In Section 4 of this paper, Hawkins shows that Weierstrass, in his theory of Elementary Divisors, had already defined a normal form equivalent to that of Jordan. Weierstrass presented his fundamental memoir "Zur Theorie der bilinearen und quadratischen Formen" to the Berlin Academy in 1868 (Monatsberichte, p. 311–338 = Werke 2, p. 19–44), two years before the publication of Jordan's Traité.

On p. 128–137 of his Traité Jordan uses his normal form to determine the set of linear substitutions commuting with a given substitution A.

According to Dieudonné (Oeuvres de Jordan, p. XIX), Jordan's research on linear substitutions was motivated by three different theories. First, in Jordan's method of constructing solvable groups (Book 4 of the Traité) linear groups appear quite naturally. Secondly, the same groups present themselves in the theory of the division of periods of Abelian functions studied by Hermite, Kronecker, and Clebsch. Finally, the studies of Mathieu on multiply transitive groups induced Jordan to study the group of projectivities of a projective line over a Galois field.

Jordan's main problem is the study of the composition of what we now call the "classical groups" over the Galois field $GF(p)$. In the following summary I shall call the groups by their modern names introduced by Dickson and modified by van der Waerden and Dieudonné, but I shall also mention some deviating names given by Jordan.

The groups studied by Jordan are:

the General Linear Group $GL(n, p)$ of all invertible linear transformations of n variables (mod p),

the Special Linear Group $SL(n, p)$ consisting of all linear transformations with determinant 1,

the corresponding projective groups $PGL(n, p)$ and $PSL(n, p)$,

the Symplectic Group $Sp(2n, p)$, called by Jordan "groupe abélien", which transforms the alternating bilinear form

$$\varphi = \sum_1^n (x_k y_{n+k} - x_{n+k} y_k)$$

into itself (modulo p),

the corresponding projective group $PSp(2n, p)$,

the "groupes de Steiner", which are affine groups of transformations transforming a quadratic form in $2n$ variables into itself (modulo 2),

orthogonal groups $O(n, p, Q)$ transforming a quadratic form Q into itself (modulo p, where p is an odd prime),

orthogonal groups $O(2n, 2, Q)$ which Jordan calls "groupes hypoabéliens", because they are contained in the "groupe abélien" $Sp(2n, 2)$.

In many of these cases, Jordan proves that these groups or their subgroups of index 2 are simple.

In 1901, L.E. Dickson published his classical treatise "Linear Groups with an Exposition of the Galois Field Theory", in which he extended Jordan's results to arbitrary Galois fields $GF(q)$. The subject was further developed by later authors, notably J.A. de Séguier and J. Dieudonné. For the history of the subject see B.L. van der Waerden: Gruppen von Linearen Transformationen (Springer-Verlag 1935, reprinted by Chelsea 1948), and J. Dieudonné: Sur les groupes classiques (Paris, Hermann 1948).

The projective group $PSL(2,p)$ can be considered as a group of fractional linear transformations in $GF(p)$

$$z' = \frac{a z + b}{c z + d}$$

with $ad - bc = 1$. This group, which is called "group of the modular equation", plays an important rôle in the theory of modular functions. It was investigated by *Galois* (Oeuvres 1897, p. 28) and by *Serret, Hermite, Mathieu*, and *Kirkman*. Its subgroups were discussed by *Betti, Hermite, Jordan*, and *Sylow*, and completely determined by *Gierster* in 1881. See the article "Endliche Gruppen" by H. Burkhardt in Enzyklopädie der math. Wissenschaften I, 1, p. 216, and my "Gruppen von linearen Transformationen", p. 8.

The symplectic groups were important for Jordan because they occur as Galois groups of the problem of the division of periods of Abelian functions having $2n$ periods. Jordan mentions this fact in his 1869 paper "Sur les équations de la division des fonctions abéliennes" in Math. Annalen I, p. 585–591 (= Oeuvres I, p 231–239). In a footnote he adds: "Nous devons à M. Kronecker la communication de cet important résultat."

Jordan's Presentation of Galois Theory

In 1869, Jordan published an exposition of Galois theory entitled "Commentaire sur Galois", Math. Annalen I, p. 142–160. With slight changes, the content of this paper was incorporated into the Traité (p. 253–274).

On p. 385–397 of the Traité the exposition is continued. Following Galois, Jordan shows that an equation can be solved by radicals if and only if its Galois group is solvable, that is, if its composition factors are prime numbers.

Next, Jordan derives another, more convenient criterion. He shows (Theorem IX on page 395):

For a group L to be solvable, it is necessary and sufficient that a subsequence of normal subgroups of L exists:

$$1 \subset F \subset G \subset H \subset ... \subset L$$

such that the substitutions of each group in the sequence are permutable modulo the preceding subgroup. In other words, it is required that the factor groups

$$F/1, \ G/F, \ H/G, \ ...$$

are all abelian.

Why is this criterion more convenient? Suppose one wants to decide whether a given group G is solvable. If one follows Galois, one has to find out whether a normal subgroup of prime index exists, and a normal subgroup of this normal subgroup, and so on. But if one follows Jordan, one has only to examine normal subgroups of the whole group G.

Geometrical Applications

Chapter 3 of Book 3 of the Traité (p. 301–333) is devoted to geometrical applications of Galois theory.

I. The first section is entitled "Equation de M. Hesse". In 1844, Ludwig Otto Hesse has proved (Gesammelte Abh., p. 123–135) that a plane cubic curve has nine inflexion points lying on twelve straight lines. If the curve is real, only three of the nine inflection points are real. Jordan denotes the nine points by the symbols

$$(0\ 0)\qquad (0\ 1)\qquad (0\ 2)$$
$$(1\ 0)\qquad (1\ 1)\qquad (1\ 2)$$
$$(2\ 0)\qquad (2\ 1)\qquad (2\ 2).$$

Nine indeterminates are introduced, which are denoted by the same symbols $(x\ y)$. The twelve lines now correspond to products

$$(x\,y)(x'\ y')(x''\ y'')$$

that satisfy the relations

$$x+x'+x''=y+y'+y''\equiv 0\qquad \text{(mod 3)}.$$

The sum of all these products is called φ. Thus we have

$$\varphi=(00)(01)(02)+(10)(11)(12)+\ldots+(02)(20)(11).$$

Jordan now proves that the Galois group of the equation on which the nine inflexion points depend, reduces itself to those substitutions of the nine indeterminates $(x\ y)$ that do not change the expression φ. This group is formed by the inhomogeneous linear transformations

$$x'\equiv a\,x+b\,y+\alpha\qquad \text{(mod 3)}$$
$$y'\equiv a'\,x+b'\,y+\alpha'\qquad \text{(mod 3)}.$$

The order of the group is

$$(3^2-1)(3^2-3)=48,$$

and the group is solvable.

In determining the Galois group, Jordan tacitly supposes the given cubic to be "generic", that is, he supposes the coefficients of its equation to be independent indeterminates. In special cases, for instance if one of the inflexion points is rational, the group may be smaller.

II. The next section is entitled "Equations de M. Clebsch". In a paper of Alfred Clebsch "Über die Anwendung der Abelschen Funktionen in der Geometrie" (Journal für Math. 63, 1864, p. 189–243) the problem was discussed:

Given a plane quartic curve, to determine a cubic curve whose intersections with the quadric are all fourfold. According to Clebsch, the problem leads to an equation X of degree

$$4^6 = 4096.$$

Jordan now determines the Galois group of the equation X. The method is similar to that applied in Section I.

In the same Section II, Jordan discussed several similar contact problems proposed by Clebsch.

III. The next problem, also due to Clebsch, is the problem of straight lines on a quartic surface having a double conic. According to Clebsch there are 16 such lines. Jordan again determines the Galois group of the problem.

IV. The problem of the 16 singular points on the "surface of Kummer" is treated by the same method. For the definition of the Kummer surface see E.E. Kummer: Über die Flächen vierten Grades mit sechzehn singulären Punkten, Monatsberichte der Berliner Akademie 1864, p. 246–260.

V. The most interesting geometrical problem discussed by Jordan is the problem of the 27 lines on a cubic surface. For the history of the subject see A. Henderson: The Twenty-seven Lines upon the Cubic Surface, Cambridge Tracts in Math. 13 (1911). The existence of these lines was discovered in a correspondence between Cayley and Salmon. Cayley found that there are lines on the surface, and Salmon found that there are (in general) just 27 lines. Our Plate 1 shows a model of a cubic surface with 27 real lines. The model itself is in Göttingen in the Mathematical Institute.

A complete description of the configuration of the 27 lines was given by Jakob Steiner in Crelle's Journal für Mathematik 53, p. 133–141 (Steiner's Werke II, p. 651–659). One of Steiner's result is: any one of the lines, say a, meets ten other lines, which form with a five triangles. Thus there are 45 triangles on the surface.

A rigorous proof of the existence of the 27 lines and the 45 triangles can be found in my "Einführung in die algebraische Geometrie" (Springer-Verlag 1939, second edition 1973) p. 148–153.

Jordan's occupation with the 27 lines begins early in 1869. In a first note in the Comptes Rendus of the Paris Academy (Oeuvres I, p. 199–202) he defines the symplectic group $Sp(2n, p)$, and he notes that $Sp(4,3)$ is just the Galois group of the equation of the 27 lines on the cubic surface. In a second note (Oeuvres I, p. 203–206) Jordan explains the relation between the symplectic group and the 27 lines in greater detail, referring for full proofs to his forthcoming Traité. In the same year 1869, he published in Journal de math. (2) 14 (Oeuvres I, p. 249–268) a paper in which the structure of the Galois group is derived by a purely combinatorial method, independent of its connection with the symplectic group. The content of this paper was incorporated into the Traité, p. 316–329.

Jordan denotes the 27 lines by single letters $a, b, \ldots$. If a is any one of the lines, the 5 triangles containing a are denoted by

$$a\,b\,c,\ a\,d\,e,\ a\,f\,g,\ a\,h\,i,\ a\,k\,l.$$

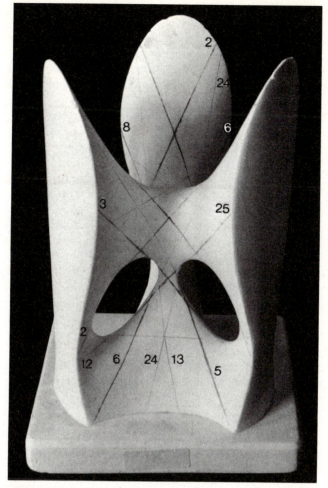

Plate 1

The 16 remaining lines that do not meet a are called, in a quite definite order:

$$m, n, p, q, r, s, t, u,$$

and

$$m', n', p', q', r', s', t', u'.$$

Now the 45 triangles can be written down as

$$abc, ade, \ldots, lps'.$$

The Galois group of the problem is certainly contained in the group of substitutions of 27 indeterminates $a, b, \ldots, u'$ transforming the function

$$\varphi = abc + ade + \ldots + lps'$$

into itself. I shall call this group G'. Its order is

$$27 \times 10 \times 8 \times 24 = 51\,840.$$

Jordan now claims that $G = G'$, and he presents arguments in favour of his assertion. In my opinion his proof is not sufficient, but the result is correct, as we shall see in the next section.

In his first note in the Comptes Rendus Jordan notes that the problem of the 27 lines is closely connected with the problem of the 28 double tangents of a quartic plane curve. I shall now explain this connection.

The 28 Double Tangents of a Plane Quartic

For the history of this subject see the article "Spezielle ebene algebraische Kurven" by G. Kohn and G. Loria in the Enzyklopädie der math. Wissenschaften III C 5, especially p. 517–542.

From the well-known Plücker formulae one easily deduces that a plane quartic curve without multiple points has just 28 double tangents. For a rigorous algebraic proof of this fact see K.G.B. Jacobi: Beweis des Satzes, dass eine Kurve n-ten Grades im allgemeinen $\frac{1}{2}n(n-2)(n^2-9)$ Doppeltangenten hat, Journal für Math. 40, p. 237–260 (1850). Our Fig. 24 shows a quartic curve having 28 real double tangents.

The first to investigate the configuration formed by the 28 double tangents and their points of contact was Jacob Steiner in 1855 (Journal für Math. 49,

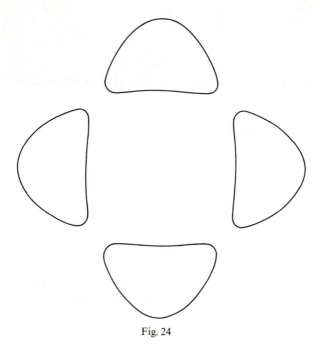

Fig. 24

p. 265–272 = Werke II, p. 605–612). He shows: if one starts with a pair of double tangents (x_1, y_1) there are 5 other pairs (x_i, y_i) $(i = 2, 3, 4, 5, 6)$ such that the 8 points of contact of any two pairs (x_i, y_i) and (x_k, y_k) always lie on a conic. Such a set of 6 pairs (x_i, y_i) is called a *Steiner complex*. There are just 63 Steiner complexes.

Three double tangents whose contact points lie on a conic form a *syzygetic triple* (from Greek syn = together and zygos = yoke). If a set of double tangents does not contain any syzygetic triple, the set is called *asyzygetic*.

Most important for the determination of the Galois group of the 28 double tangents is a generation of the quartic curve discovered by Aronhold (Monatsberichte der Berliner Akademie 1864, p. 499–523). Aronhold's method may be explained as follows.

Seven given points in the plane determine, in general, a linear set of cubic curves passing through the seven points. Any two curves of the set intersect in two more points. If the two points coincide, the curves have a common tangent at that point (see Fig. 25). These ∞^1 common tangents form a "curve of class 4", that is, the dual of a quartic curve.

Now consider the dual situation. Instead of the cubic curves passing through seven given points we may consider, with Aronhold, the linear set of curves of class 3 containing seven given lines. Every pair of these curves has two more lines in common. If they coincide, the two dual curves have a point of contact, and these points of contact lie on a quartic curve.

Aronhold shows that the seven given lines form a maximal asyzygetic set of double tangents of this quartic, and that the other 21 double tangents can be constructed rationally from the seven. Every quartic curve without double

Fig. 25

points can be obtained by this construction, and every asyzygetic set of seven double tangents can be used to generate the quartic. For rigourous proofs of these statements see H. Weber: Lehrbuch der Algebra II, second edition, p. 425–447.

Now it is easy to determine the Galois group P of the equation determining the 28 double tangents. One starts with a "generic" set of seven lines, that is, one assumes the inhomogeneous coordinates of these lines to be independent indeterminates. Now one constructs the curve; it will be a generic quartic curve. There are

$$8 \times 36 = 288$$

asyzygetic sets of seven lines, and in each of the sets there are 7! ways of numbering the lines. Any one of the numbered sets can be replaced by any other of the

$$288 \times 7! = 1\,451\,520$$

numbered sets. Every such replacement yields an automorphism of the field of rational functions of the coordinates of the seven lines, and these isomorphisms leave the quartic invariant. These automorphisms form the Galois group P.

I have used the modern expressions "automorphism" and "generic", but the same ideas can also be expressed in the terminology of Galois and Jordan: see again Weber's Lehrbuch, p. 447–454.

On p. 454–458 Weber proves that the group P is simple.

The relation between the 28 double tangents and the 27 lines on a cubic surface was established by M. Geiser in 1868. His paper "Über die Doppeltangenten einer ebenen Kurve vierten Grades" was published in Math. Annalen I, p. 129–138. Geiser's method can be explained as follows.

From an arbitrary point A on the cubic surface, not lying on one of the 27 lines, one draws all tangents to the surface. Apart from the tangents at A, which form the tangential plane, all tangents lie on a quartic cone. The intersection of the cone with an arbitrary plane π is a quartic curve. The tangent plane at A intersects π in a double tangent of the quartic. The other 27 double tangents lie in the planes connecting A with the 27 lines on the cubic surface.

The Galois group of the 27 lines can now be obtained from that of the 28 double tangents as follows. In the latter group, considered as a group of permutations of the double tangents, take the subgroup that leaves invariant the double tangent lying in the tangential plane. This subgroup permutes the 27 other double tangents, and hence it induces a group of permutations of the 27 lines on the cubic surface. This group G is the Galois group of the 27 lines. Its order is, obviously,

$$\frac{1\,451\,520}{28} = 51\,840.$$

As we have seen, Jordan constructed a group G' of just this order, and he proved that the Galois group G is contained in G'. Since the orders are equal, Jordan's assertion $G = G'$ is justified.

On p. 329-333 Jordan investigates the Galois group of the 28 double tangents by another method. He considers the problem of Clebsch: To find all curves of order $n-3$ having $\frac{1}{2}n(n-3)$ points of contact with a given curve of order n. The method of Clebsch uses Abelian functions. For $n=4$ one has the problem of finding the double tangents of a quartic curve.

Application of Galois Theory to Transcendental Functions

In Chapter 4 of Book 3 of his Traité, Jordan applies the theory of Galois to problems concerning transcendental functions.

Let me first remind the reader that Jordan makes a distinction between the *algebraic Galois group* of an equation

(1) $$f(z, w) = 0$$

and its *monodromy group* with respect to the complex variable z. Let the coefficients of the equation (1) be taken from a field of constants K, which may also contain variable parameters. In forming the algebraic Galois group of the equation (1), one takes as a ground field the field $K(z)$ of rational functions of z with coefficients in K. The algebraic group is a group of permutations of the roots $w_1, \ldots, w_n$. It contains as a subgroup the monodromy group defined by Puiseux and Hermite. If the field of constants K is enlarged by the adjunction of certain algebraic elements, the algebraic group is reduced to the monodromy group.

I shall now summarize the single sections of Jordan's Chapter 4 (p. 334–382).

I. Jordan first considers the problem of determining $\cos(x/n)$ if $\cos x$ is given. The quantity $\cos(x/n)$ is linked to $\cos x$ by an equation of degree n, whose roots are

$$(p) = \cos\frac{x + 2p\pi}{n}$$

and the monodromy group of the equation is formed by the substitutions

$$p' \equiv p + m \quad (\bmod\ n).$$

This is also the Galois group of the equation after the adjunction of $\cos(2\pi/n)$ to the ground field. If the ground field is the field of rational numbers, the substitutions of the algebraic Galois group are of the form

$$p' \equiv ap + b \quad (\bmod\ n).$$

II. Jordan now passes to the theory of elliptic functions.
Let $u = \lambda(z)$ be the inverse function of

$$z = \int_{\infty}^{u} \frac{dv}{\sqrt{(1 - v^2)(1 - k^2 v^2)}}.$$

Its derivative is

$$\lambda'(z) = \sqrt{(1 - \lambda^2(z))(1 - k^2 \lambda^2(z))}.$$

Now if $\lambda(z)$ and $\lambda'(z)$ are given, $\lambda(z/n)$ is a root of an equation of degree n^2. The roots of this equation are

$$(p\,q) = \lambda \left(\frac{z + p\,\omega + q\,\omega'}{n} \right)$$

where ω and ω' are the fundamental periods of the elliptic function $\lambda(z)$.

Supposing n to the prime, Jordan shows that the substitutions of the Galois group of the equation are all of the form

(2)
$$\begin{aligned} p' &\equiv a\,p + b\,q + m \qquad (\text{mod } n) \\ q' &\equiv a'\,p + b'\,q + m' \qquad (\text{mod } n). \end{aligned}$$

If one adjoins to the field K the constants

$$\lambda(\omega/n),\ \lambda'(\omega/n),\ \lambda(\omega'/n),\ \lambda'(\omega'/n),$$

the Galois group will be reduced to the monodromy group

$$\begin{aligned} p' &\equiv p + m \qquad (\text{mod } n) \\ q' &\equiv q + m' \qquad (\text{mod } n). \end{aligned}$$

Since this group is abelian, the resolution of the equation for $\lambda(z/n)$ offers no problem. Jordan's main problem is: to find the Galois group of the equation of degree n^2 determining

$$\lambda(\omega/n) \quad \text{and} \quad \lambda'(\omega/n)$$

where ω is any primitive period. The roots of this equation are all of the form

$$(p\,q) = \lambda \left(\frac{p\,\omega + q\,\omega'}{n} \right).$$

Eliminating the root $\lambda(0/n)$, one is left with an equation of degree $n^2 - 1$. The monodromy group of this equation with respect to the complex variable k consists of linear transformations of the form

$$\begin{aligned} p^* &\equiv a\,p + b\,q \qquad (\text{mod } n) \\ q^* &\equiv a'\,p + b'\,q \qquad (\text{mod } n) \end{aligned}$$

with $a\,b' - a'\,b \equiv 1$ (mod n). These transformations form the group $SL(2, n)$.

III. In the preceding Sections I and II, the functions under consideration were the inverse functions of Abelian integrals on algebraic curves of genus 0 or 1. Now Jordan passes to the case of genus 2. In this case we have hyperelliptic integrals

$$z = \int_0^u \frac{\mu + v\,v}{\Delta(v)}\,dv$$

where $\Delta(v)$ is a polynomial of degree 6, and the inverse functions are Abelian functions of two complex variables having 4 periods. Following Hermite, Jordan investigates the division problem of these functions, and he ends up with the symplectic group $Sp(4, p)$.

In the case $p = 3$ he finds that the group $PSp(4,3)$ is isomorphic to the simple group G of the 27 lines on a cubic surface.

IV. In the last section of Book 3 Jordan investigates the possibility of solving equations by means of transcendental functions. In particular he discusses the methods of Hermite, Kronecker, and Brioschi for solving equations of degree 5 by means of modular functions and elliptic functions. For higher degrees, Jordan shows that such a solution is impossible.

On Solvable Groups

The fourth book of the Traité is devoted to the problem: to construct, for any given degree d, all solvable transitive groups of substitutions on d letters. This is what Jordan calls Problem A.

Jordan shows that Problem A for non-primitive groups can be reduced to the same problem for primitive groups, and that primitive groups necessarily have degree $d = p^n$, where p is prime.

Galois has solved Problem A for groups of degree p, and he has found some partial results for primitive groups of degree p^2. Jordan now undertakes to solve Problem A for primitive groups of degree p^n.

A group of linear transformations of variables $x_1, \ldots, x_n \pmod p$ is called in modern terminology *irreducible*, and in Jordan's terminology *primaire*, if it is not possible to find linear functions $y_1, \ldots, y_m$ of the x_i, in number less than n, which are transformed by the group into linear functions of themselves.

Jordan now shows that Problem A can be reduced to Problem B:

B. To construct the maximal solvable irreducible groups contained in the linear group $GL(n, p)$.

A special case of Problem B is Problem C:

C. To construct the maximal solvable irreducible groups contained in the symplectic group $Sp(2n, p)$ or in one of the hypoabelian groups $O^+(2n, 2)$ or $O^+(2n+1, 2)$.

The pages 410–662 are devoted to the solution of the Problems B and C. Jordan's method is recursive. He indicates a method for solving his problems for groups of degree p^n, supposing that they are solved for lower degrees p^m $(m < n)$. He finally arrives at a complete classification of the groups in question.

In a sequence of papers, published between 1871 and 1875 and republished in Oeuvres I, p. 277–495, Jordan has completed and extended the results obtained in the Traité. His last two papers on the construction and classification of solvable permutation groups, published in 1908 and 1917 and republished as papers (121) and (126) in Oeuvres II, have been summarized by J. Dieudonné in his introduction to Oeuvres I, p. XXXIV–XLI.

Part Two
Groups

Chapter 8
Early Group Theory

What is early, what is late? Be it sufficient to say that the present chapter deals mainly with the nineteenth century, but if the subject matter requires it, I shall extend my account to the first half of the twentieth century.

After the appearance of Jordan's Traité in 1870, a fundamental change of character of the theory of groups took place. Before 1870, only two kinds of groups were considered, namely groups of substitutions (or permutations) and groups of geometrical transformations. After 1870, the abstract notion "group" was developed in several steps, notably by Kronecker (1870), Cayley (1878), von Dyck (1882), and Weber (1882). The modern definition of a group by means of axioms was given for abelian groups by Kronecker (1870), for finite groups by Weber (1882), and for infinite groups by the same Weber (1893). The whole development of these notions has been described in great detail by H. Wussing in his book "Die Genesis des abstrakten Gruppenbegriffs" (1969).

After the introduction of the abstract notion "group" the main problem of group theory was: to investigate the structure of groups independent of their representation by permutations or transformations, and only afterwards to study these representations.

Accordingly, the present chapter will be divided into four parts:
A. Groups of Substitutions
B. Groups of Transformations
C. Abstract Groups
D. The Structure of Finite Groups

Part A
Groups of Substitutions

Early Theorems Concerning Subgroups of S_n

The earliest authors who investigated groups of substitutions were Lagrange, Ruffini, and Cauchy. They were mainly concerned with "*generic*" (or "general") equations of a given degree n, that is, with equations in which either the coefficients or the roots $x_1, \ldots, x_n$ are independent variables. They were interested in forming *auxiliary equations*, if possible of lower degrees. For this purpose, they considered rational functions

$$f(x_1, \ldots, x_n)$$

and they asked: How many different values does the function f assume if the roots are permuted? If $f_1, \ldots, f_s$ are these values, they will be the roots of an auxiliary equation of degree s:

$$(t - f_1)(t - f_2) \ldots (t - f_s) = 0.$$

For this reason, our early authors were very much interested in the possible values of s.

If H is the subgroup of S_n that transforms f into itself, the cosets of H transform f into $f_1, \ldots, f_s$. So the number s is just *the index of H in S_n*.

For instance, if $n = 4$ and f is the function

$$f = x_1 x_2 + x_3 x_4,$$

the index s is 3, and f is a root of a cubic equation. This was noted already by Lagrange, and it was the basis of his solution of the quartic equation.

The case $n = 4$, in which an auxiliary equation of lower degree can be used to solve the original equation, is an exception. If n exceeds 4, the index s is either 2 or at least n. This was proved by Cauchy in 1845 (Comptes Rendus Acad. Paris 21, p. 1101).

In S_5 there is a subgroup L of index $s = 6$ and order 20, namely the group of linear index substitutions

$$k' \equiv a k + b \quad (\text{mod } 5).$$

If f is a function invariant only under this subgroup, it takes six different values if the roots are permuted, and these values are the roots of an auxiliary equation of degree 6. The permutations of S_5 induce permutations of the six values, hence S_5 is isomorphic to a transitive subgroup H of index 6 in S_6.

This subgroup H is again an exception. For $n > 6$ every subgroup of index n in S_n leaves one of the x_k invariant. This was proved by J.A. Serret in a paper entitled "Mémoire sur le nombre de valeurs que peut prendre une fonction quand on y permute les lettres qu'elle enferme", Journal de Math. (1) 15, p. 1–44.

In the same paper Serret proves:

If $s > n$ and $n > 8$,   then $s \geq 2n$.

Later authors have shown that this is true for all $n > 6$. Serret also proves:

If $s > 2n$,   then $s \geq 1/2 \ n(n-1)$.

More results concerning the possible value of the index s were obtained by *Mathieu, Jordan, Sylow, Netto, Frobenius, Borchert, Maillet,* and *Miller.* See the article of H. Burkhardt in Enzyklopädie der math. Wissenschaften I, 1, p. 213 f.

Mathieu

Two very interesting papers of E. Mathieu on multiply transitive groups of substitutions were published in 1861 and 1873 respectively. The first paper is entitled "Mémoire sur l'étude des fonctions de plusieurs quantités, sur la manière de les former et sur les substitutions qui les laissent invariables", Journal de Math. (2) 6, p. 241–323 (1861).

In this paper Mathieu explains a general method to obtain multiply transitive groups. In particular he constructs a five-fold transitive group of substitutions on 12 letters. In the same paper he announces the existence of a five-fold transitive group of permutations of 24 letters. The construction of this group is described in a second paper entitled "Sur la fonction cinq fois transitive de 24 quantités", Journal de Math. (2) 18, p. 25–46 (1873).

The starting point of Mathieu's constructions is the projective linear group of fractional linear transformations

$$z' = \frac{a z + b}{c z + d}$$

with coefficients from a Galois field $K = GF(q)$. This group is treefold transitive on the projective line over K. Starting with this group, Mathieu succeeds in constructing his five-fold transitive groups, by ingenious artifices.

Sylow

In the spring of 1872 the Norwegian mathematician M.L. Sylow presented to the Mathematische Annalen a paper of fundamental importance for the structure theory of finite groups, entitled "Théorèmes sur les groupes de substitutions".

This paper, published in Math. Ann. 5, p. 584–594, contains full proofs of eight theorems. Most interesting are the following four:

Theorem I. If p^α is the largest power of the prime p contained in the order of the group G, there is a subgroup H of order p^α. If the normalizer of H is of order $p^\alpha m$, the order of G is

$$p^\alpha m (p r + 1).$$

Theorem II. The group G contains exactly $p r + 1$ subgroups of order p^α. One obtains them by transforming one of them by the substitutions of G, each of the subgroups being produced by $p^\alpha m$ different transformants.

Theorem III. If the order of a group G is p^α, p being prime, every substitution θ of the group can be expressed by the formula

$$\theta = \theta_0^i\, \theta_1^k\, \theta_2^l \ldots \theta_{\alpha-1}^r$$

with

$$\theta_0^p = 1$$
$$\theta_1^p = \theta_0^a$$
$$\theta_2^p = \theta_0^b\, \theta_1^c$$
$$\theta_3^p = \theta_0^d\, \theta_1^e\, \theta_2^f$$
$$\ldots$$

and

$$\theta^{-1}\theta_0\,\theta = \theta_0$$
$$\theta^{-1}\theta_1\,\theta = \theta_0^\beta\,\theta_1$$
$$\theta^{-1}\theta_2\,\theta = \theta_0^\gamma\,\theta_1^\delta\,\theta_2$$
$$\ldots .$$

Note that θ_0 is in the centre and generates a subgroup of order p. Modulo this subgroup θ_1 is again in the centre, and so on. It follows that groups of order p^α are solvable.

As a corollary, Sylow states:

If the order of the Galois group of an algebraic equation is a power of a prime, the equation is solvable by radicals.

A more detailed analysis for the case of a transitive group H yields

Theorem IV. If the degree of an irreducible equation is p^β and if the order of its group is also a power of the prime p, each one of its roots can be obtained by a sequence of β abelian equations of degree p.

Sylow notes that the special case $p=2$ of this theorem has already be obtained in 1871 by M.J. Petersen.

In 1887, G. Frobenius published a new proof of Sylow's theorems. His paper in Crelle's Journal für Math. 100, p. 179–181, is entitled "Neuer Beweis des Sylowschen Satzes". He first notes that every finite group can be represented as a group of substitutions of its own elements, so that Sylow's proof is valid for abstract groups as well, but he does not want to use this representation. Following Weber, Frobenius defines an abstract finite group by four postulates, and he presents a new proof of Sylow's theorems based on these postulates.

Part B
Groups of Transformations

As we have seen at the beginning of Chapter 7, Leonard Euler and Olinde Rodrigues analysed the structure of the group of rigid motions in 3-space, and Camille Jordan undertook a systematic investigation of the closed subgroups of this group. It seems that Jordan was the first to use the word "groupe" for groups of geometrical transformations.

Next we have to discuss the investigations of Arthur Cayley and Felix Klein on non-Euclidean geometry, of Felix Klein on discrete groups of fractional linear transformations, and of Sophus Lie on continuous groups.

Non-Euclidean Geometry

On the work of Arthur Cayley see the excellent article by John D. North in the Dictionary of Scientific Biography. Cayley's work in all parts of algebra and geometry has been extremely influencial. In a sequence of papers entitled "Memoirs upon Quantics", written between 1854 and 1878, Cayley laid the foundations of the Theory of Invariants.

In his "Sixth Memoir upon Quantics", published in 1859 in Vol. 149 of the Philos. Transactions of the Royal Society, Cayley first develops the projective geometry of points, lines, and conics in the projective plane, starting with coordinates (x, y, z) of points and (ξ, η, ζ) of lines. I shall denote these coordinates by (x_1, x_2, x_3) and (u_1, u_2, u_3). A quadratic form

$$f(x, x) = \Sigma \, a_{ik} \, x_i \, x_k \qquad (a_{ik} = a_{ki})$$

defines a conic. The polar form

$$f(x, y) = \Sigma \, a_{ik} \, x_i \, y_k$$

defines the polar of a point y. The covariant form

$$F(u, u) = \Sigma \, A_{ik} \, u_i \, u_k$$

in which the A_{ik} are proportional to the subdeterminants of the matrix (a_{ik}), defines the "dual conic" formed by the tangents of the original conic. One can also start with a dual conic F and form the covariant f by means of subdeterminants.

Cayley next develops a theory of "distances" by assuming a fixed conic which he calls "the absolute". He defines the "distance" between two points x and y by

(1)
$$\text{Dist}(x, y) = \cos^{-1} \frac{f(x, y)}{\sqrt{f(x, x)} \, \sqrt{f(y, y)}}$$

and the "distance" or angle between two lines by

(2)
$$\text{Dist}(u, v) = \cos^{-1} \frac{F(u, v)}{\sqrt{F(u, u)} \, \sqrt{F(v, v)}}.$$

Cayley shows that $\text{Dist}(x, y)$ is the integral of an infinitesimal distance ds, taken over a straight path of integration from x to y. If y lies on the path of integration from x to z, it follows that

(3) $$\text{Dist}(x, y) + \text{Dist}(y, z) = \text{Dist}(x, z).$$

Cayley considers two special cases. In the first case the form is positive definite and can be written as

$$f(x, x) = x_1^2 + x_2^2 + x_3^2.$$

In this case we have

$$F(u, u) = u_1^2 + u_2^2 + u_3^2$$

and

(4) $$\text{Dist}(x, y) = \cos^{-1} \frac{x_1 y_1 + x_2 y_2 + x_3 y_3}{\sqrt{x_1^2 + x_2^2 + x_3^2} \sqrt{y_1^2 + y_2^2 + y_3^2}}$$

(5) $$\text{Dist}(u, v) = \cos^{-1} \frac{u_1 v_1 + u_2 v_2 + u_3 v_3}{\sqrt{u_1^2 + u_2^2 + u_3^2} \sqrt{v_1^2 + v_2^2 + v_3^2}}.$$

The lines and planes passing through a fixed point O in Euclidean 3-space can be considered as "points" and "lines" in a projective "plane", and (4) is the well known formula for the angle between two vectors in 3-space. It follows that the "distances" between the "points" and "lines" in the projective "plane" just defined are the angles between lines and planes passing through O. These "points" and "lines" form, in the terminology of Felix Klein, an "elliptic plane".

The second case considered by Cayley is a limiting case. One obtains it by starting with the positive forms

$$f(x, x) = \varepsilon x_1^2 + \varepsilon x_2^2 + x_3^2$$
$$F(u, u) = u_1^2 + u_2^2 + \varepsilon u_3^2$$

and letting ε go to zero. One obtains in the limit

$$f(x, x) = x_3^2$$
$$F(u, u) = u_1^2 + u_2^2.$$

The conic $f = 0$ is the line at infinity, counted twice. The dual conic $F = 0$ is a pair of pencils of lines defined by the points at infinity $(1, i, 0)$ and $(1, -i, 0)$. These points are just the "points circulaires" of Poncelet, so called because they are common to all circles in the Euclidean plane. One obtains in the limit, introducing inhomogeneous rectangular coordinates x and y,

(6) $$\text{Dist}(P, P') = \sqrt{(x - x')^2 + (y - y')^2}$$

and

(7)
$$\text{Dist}(u, v) = \cos^{-1} \frac{u_1 v_1 + u_2 v_2}{\sqrt{u_1^2 + u_2^2} \cdot \sqrt{v_1^2 + v_2^2}},$$

that is, one gets the Euclidean distance between two points and the Euclidean angle between two lines.

What we call "projective geometry" is called by Cayley "descriptive geometry". According to Cayley, one passes from projective geometry to metrical geometry by fixing a conic and calling it "the absolute". He concludes:

Metrical geometry is thus a part of descriptive geometry, and descriptive geometry is *all* geometry and reciprocally …

Today we do not share this restricted view of geometry. For us, projective geometry is not "all geometry". For instance, topology and Riemannian geometry are not parts of Cayley's "descriptive geometry".

Felix Klein, in his two papers "Über die sogenannte Nicht-Euklidische Geometrie", in Math. Annalen 4 (1871), p. 573–625 and Math. Annalen 6 (1873), p. 112–145, distinguishes two types of Non-Euclidean geometry, which he calls "elliptic" and "hyperbolic". In the elliptic case the quadratic form $f(x, x)$ defining the absolute conic is positive definite: the conic has no real points, and we can use Cayley's formulae (4) and (5). Klein's "elliptic plane" can be obtained from a sphere in Euclidean 3-space by identifying opposite points.

Besides this elliptic plane and the Euclidean plane considered by Cayley, Klein investigates a third case in which the absolute conic is a non-singular conic having real points. As a model, one can take a circle in the Euclidean plane. The forms defining the conic and its dual conic are now

(8)
$$f(x, x) = -x_1^2 - x_2^2 + x_3^2$$

(9)
$$F(u, u) = u_1^2 + u_2^2 - u_3^2.$$

Klein's model of the "hyperbolic geometry" of Lobatchewsky and Bolyai is obtained by restricting oneself to inner points of the circle:

$$f(x, x) > 0,$$

and to lines containing inner points:

$$F(u, u) > 0.$$

In this case formula (2) defining the "distance" or angle between two lines can be retained, but (1) has to be modified. If x and y are inner points of the circle, the intersections of the connecting line with the circle can be obtained from a quadratic equation

$$\lambda_1^2 f(x, x) + 2 \lambda_1 \lambda_2 f(x, y) + \lambda_2^2 f(y, y) = 0.$$

The equation has two real roots, hence we have

$$f(x, y)^2 - f(x, x) f(y, y) > 0.$$

So the argument of the function $\cos^{-1}$ in (1) is larger than 1, and the arc cosine is purely imaginary.

Felix Klein now replaces the arc cosine by the logarithm of a cross-ratio. If P and Q are the points formerly called x and y, and if A and B are the points of intersection of PQ with the conic, the cross-ration of A and B with respect to P and Q is a projective invariant, and its logarithm is Klein's "distance".

By the way, Cayley's "distance" (1) is just equal to $\pm i/2$ times Klein's distance.

When Volume 2 of Cayley's Collected Papers appeared in 1889, Cayley added a note to his "sixth memoir", saying that Klein's replacement of the arc cosine by a logarithm is a great improvement.

In Klein's first paper on non-Euclidean geometry no groups are considered, but in the second paper of 1873 the notion of "transformation group" occurs. Klein considers invertible transformations of a manifold, and he defines the notion "group", as in Jordan's Traité, by the property: if A and B are in the group, so is AB. Later on, Klein has seen that it is necessary to require that A^{-1} is in the group if A is.

In §3 of Klein's second paper on non-Euclidean geometry in Math. Annalen 6, Klein introduces a group he calls "Hauptgruppe". It is generated by the Euclidean displacements, the similarity transformations, and the reflexions.

In §4 Klein explains that each one of the different "methods of geometry" is characterized by a group of transformations. This is also the fundamental idea in Klein's famous "Erlanger Programm" (1872). On the history of this "program" see David E. Rowe: A Forgotten Chapter in the History of Felix Klein's Erlanger Programm, Historia Mathematica 10, p. 448–457 (1983).

According to Klein, projective geometry deals with those properties of figures that are invariant under projective transformations, Euclidean geometry with properties invariant under the "Hauptgruppe", and so on. The groups of the elliptic and hyperbolic geometries are the groups of projectivities transforming a conic (or in three dimensions a quadratic surface) into itself.

Felix Klein and Sophus Lie

Klein and Lie became friends at Berlin. In 1870 they went to Paris, where they lived in adjacent rooms for two months. As we have seen in Chapter 4, they were very much impressed by Camille Jordan, whose Traité des substitutions had just appeared.

In 1871 a joint paper of Klein and Lie entitled "Über diejenigen ebenen Kurven, welche durch ein geschlossenes System von einfach unendlich vielen vertauschbaren linearen Transformationen in sich übergehen" was published in Math. Annalen 4, p. 424–429.

The expression "geschlossenes System von linearen Transformationen" in the title just means "group of linear transformations". In fact, the authors state:

The expression "closed system of transformations" corresponds to what in the theory of substitutions is denoted by the term "group of substitutions".

The groups considered by Klein and Lie are one-dimensional continuous groups. In a footnote the authors state correctly that all one-dimensional continuous groups are commutative. In the same footnote the authors give an example of a three-dimensional continuous group which is not commutative, namely the group of projectivities of a plane that transform a conic into itself. From this paper one sees that Lie's ideas about continuous groups of transformations began to take shape about 1870.

In later years, the ideas of the two friends went into different directions. Lie developed his theory of continuous groups and applied it to the study of differential equations, whereas Klein mainly investigated discrete groups. Discrete groups of fractional linear transformations play an important role in the study of automorphic functions.

The testimonies of Klein and Lie on their early friendship and their later divergent development are reproduced on p. 153 of the book of H. Wussing: Die Genesis des abstrakten Gruppenbegriffs (Verlag der Wissenschaften, East-Berlin 1969).

On the further development of the theory of discrete groups and automorphic functions see the article "Automorphe Funktionen" by R. Fricke in Encyclopädie der mathematischen Wissenschaften II B 4 (1913), and also my report "Gruppen von linearen Transformationen", Ergebnisse der Mathematik IV, 2 (Springer 1935).

Felix Klein on Finite Groups of Fractional Linear Transformations

In June 1875, Felix Klein submitted to the Math. Annalen an important paper entitled "Über binäre Formen mit linearen Transformationen in sich selbst" (Math. Annalen 9, p. 209–217). In this paper he determined all finite groups of fractional linear transformations of a complex variable z:

$$(10) \qquad z' = \frac{a z + b}{c z + d}.$$

These transformations transform circles into circles, while preserving the orientation of the function-theoretical z-plane. If the points and circles in the plane are transferred to a sphere by means of a stereographic projection, one obtains transformations of the sphere into itself preserving the orientation, which can be extended to projective transformations of the real projective space P_3 into itself. These projectivities are, in the terminology of Klein, *hyperbolic motions*. Conversely, every orientation-preserving projective transformation that transforms the sphere into itself yields a fractional linear transformation of the complex variable z. So the problem of determining all finite groups of transformations (10) is equivalent to the problem of finding all finite groups of hyperbolic motions in P_3.

Klein first notes: if a hyperbolic motion has a finite order, it must be a "rotation" leaving invariant all points of an axis connecting two real points on the sphere. Next he proves: if the product of two rotations is again a rotation, the axes must intersect in a point inside the sphere. From this he concludes: all axes of rotations belonging to the finite group G intersect in one point, which is a fixed point of the group.

We may assume, Klein says, that this fixed point is the centre of the sphere. In this case the rotations are Euclidean rotations. Now the finite groups of Euclidean rotations are known from a classical investigation of Jordan (see Chapter 7). They are:

1) Cyclic groups,
2) Dihedral groups,
3) the tetrahedral group,
4) the octahedral group,
5) the icosahedral group.

Thus, Klein's problem is completely solved.

A direct derivation of the types of finite groups of fractional linear transformations was given by H.H. Mitchell in 1911 (Transactions Amer. Math. Soc. 12, p. 208–211).

The icosahedral group is isomorphic to the alternating group A_5. Hence the icosahedron can be used to illustrate the Galois theory of the quintic equation. This was done in Klein's very nice booklet "Vorlesungen über das Ikosaeder und die Auflösung der Gleichungen vom fünften Grad" (Leipzig 1884).

Sophus Lie

The Norwegian Sophus Lie, the founder of the theory of "Lie Groups", was born at Nordfjordeid in December 1842. An account of his life and work was given by H. Freudenthal in the Dictionary of Scientific Biography, Vol. 8, p. 323–327.

According to Lie's own biographical statements, his friendship with Felix Klein, whom he met at Berlin in the winter 1869/70, was of great importance for his later work on groups of transformations. In the summer of 1870, Lie discovered his famous contact transformation, which transforms straight lines into spheres (see Sophus Lie: Gesammelte Abhandlungen, Vol. 1, p. 93–96). In 1871 Lie obtained a scholarship from the university of Christiania, and in 1872 he finished his PhD-thesis entitled "Over en Classe geometriske Transformationer" (Gesammelte Abhandlungen, Vol. 1, p. 105–214, in German).

During this time Lie developed his integration theory of partial differential equations. According to Freudenthal, this theory is "now found in many textbooks, although rarely under his name". For a survey of this theory see Freudenthal's article in D.Sc.B. mentioned before.

Lie's investigations on the integration of differential equations induced him to consider groups of transformations transforming a differential equation into itself. It is true that these investigations were published only much later (1882–1883) in a sequence of papers (Gesammelte Abhandlungen 5, p. 238–313 and 362–424), but we know from Engel's introduction to this volume 5 that Lie's occupation with this subject began at least ten years earlier. Engel informs us that a treatise on differential equations, which Lie intended to write, has remained unwritten. Engel continues (Vol. 5, p. VIII):

> This is a pity, for Lie attached great weight to just these applications. Originally he had developed his whole theory of transformation groups only because this theory was the instrument he needed to treat his integration problems.

In the present chapter, I only want to discuss the pre-history of Lie's theory of finite-dimensional continuous groups. The theory itself will be discussed in Chapter 9.

Part C
Abstract Groups

The subject matter of the present Part C has been treated more fully by H. Wussing: Die Genesis des abstrakten Gruppenbegriffes, East-Berlin 1969.

Abstract algebraic structures, defined solely by the laws of composition of theirs elements, occur already at an early stage. For instance, the rules for adding and multiplying complex numbers $a \pm bi$ were explained as early as 1560 in Bombelli's "Algebra".

As we have seen in Chapter 6, Galois defined his "Galois Fields" $GF(q)$ by describing the laws of composition of the elements. The same holds for Hamilton, who discovered the algebra of quaternions in October 1843, as we shall see in Chapter 10.

It is very remarkable that abstract fields and algebras were discovered at such an early stage. In the case of groups the history took a different course. Galois introduced groups of substitutions in 1829, and Jordan investigated groups of motion in 1867. The first to introduce abstract *abelian* groups, defined by the rule of composition of their elements, was Kronecker in 1870. The first steps towards the *general* notion of an abstract group was taken by Cayley in 1854 and 1878, and the first clear definition of this notion was given by Walter van Dyck in 1882. In the same year, Heinrich Weber presented another, equivalent definition.

I shall now describe the development in greater detail, starting with the work of Euler and Gauss on number theory.

Leonhard Euler

In a paper entitled "Theoremata circa residua ex divisione potestatum relicta" (Theorems on the residues left by the division of powers), published in 1761 (Opera omnia, series prima, Vol. 2), Euler divides the powers a^μ of any integer a by a prime p, and considers the remainders of this division. If a is not divisible by p, there is a power a^λ which yields residue 1, and if λ is the least integer having this property, the powers

$$1, a, a^2, \ldots, a^{\lambda-1}$$

yield just λ different residues. Euler now proves that λ is a divisor of $p-1$. The method is the same by which Lagrange proved that the order of any element in a group of substitutions is a divisor of the order of the group, and by which Jordan proved that the order of any subgroup of a finite group is a divisor of the order of the group, namely: the set of all non-zero residues modulo p is divided into cosets

$$(r, a\,r, a^2\,r, \ldots, a^{\lambda-1}\,r).$$

As an application, Euler presents a "more natural" proof of a theorem of Fermat

$$a^{p-1} \equiv 1 \qquad (\mathrm{mod}\,p)$$

which he had proved earlier by means of the expansion of $(a+b)^p$.

Carl Friedrich Gauss

In the last part of his "Disquisitiones arithmeticae", Gauss defines a "composition" of binary quadratic forms as follows.
    If a form

$$F = AX^2 + 2BXY + CY^2$$

can be transformed into a product of two forms

$$f = ax^2 + 2bxy + cy^2, \qquad f' = a'x'^2 + 2b'x'y' + c'y'^2$$

by a substitution

$$X = p\,x\,x' + p'\,x\,y' + p''\,y\,x' + p'''\,y\,y'$$
$$Y = q\,x\,x' + q'\,x\,y' + q''\,y\,x' + q'''\,y\,y',$$

then Gauss says that F is *transformable* into ff', and if the six integers

$$p\,q' - q\,p', \; p\,q'' - q\,p'', \ldots, p''\,q''' - q''\,p'''$$

have no common factor, he says that F is *composed* from the forms f and f'.
    The form f is called *primitive*, if a, b, c have no common factor. If the primitive forms f and f' have the same discriminant D, it is always possible to find a form F which has the same discriminant, and is composed of f and f'.

Two forms f and g are defined to be in the same class, if f can be transformed into g by a linear transformation

$$x = \alpha x' + \beta y'$$
$$y = \gamma x' + \delta y'$$

having determinant 1. Gauss proves: if F is composed from f and f', the class of F is uniquely determined by the classes of f and f'. So the composition of forms yields a composition of classes, which is commutative and associative. In modern terminology, the classes of primitive forms having a given discriminant form a *finite abelian group*. The unit element of the group is represented by the *principal form*

$$x^2 - D y^2.$$

Ernst Schering

Ernst Schering, a pupil of Gauss, investigated the structure of the group of classes of binary quadratic forms of a given discriminant D. His paper "Die Fundamental-Classen der zusammensetzbaren arithmetischen Formen" was published in 1869 in the Abhandlungen of the Göttingen academy 14, p. 3–13. In this paper he proved that the group of quadratic forms having a given discriminant possesses a set of generators $A, B, C, \ldots$ of orders $a, b, c, \ldots$ such that every class can be uniquely expressed as a product

$$A^\alpha B^\beta C^\gamma \ldots$$

in which α runs over the residue classes modulo a, and β over the residue classes modulo b, etc. This is what we now call the "fundamental theorem on finite abelian groups".

Implicitly, the same result had already been obtained by Abel in his paper on equations having a commutative Galois group (Oeuvres I, p. 499). As we shall see presently, the "fundamental theorem" holds for every finite abelian group.

Leopold Kronecker

The first German mathematician who fully realized the importance of the investigations of Abel and Galois on the solvability of algebraic equations was Leopold Kronecker. In 1853 he published a paper "Über die algebraisch auflösbaren Gleichungen, erste Abhandlung", Bericht über die Verhandlungen der Akademie Berlin 1853, p. 365–374. In this paper he announces the important theorem:

"The roots of every abelian equation with integer coefficients can be represented as rational functions of roots of unity."

As we have seen in Chapter 7, Kronecker also investigated the Galois group of the problem of the division of abelian functions having $2n$ periods. He found that this Galois group is just the symplectic group $Sp(2n, p)$, and he informed Jordan of this fact.

In 1870, Kronecker published a paper entitled "Auseinandersetzung einiger Eigenschaften der Klassenzahl idealer komplexer Zahlen", Monatshefte der Berliner Akademie 1870, p. 881–889. In this paper, Kronecker introduces the notion of an *abstract abelian group*. He considers a finite number of elements

$$\theta', \theta'', \dots$$

such that from any two of them a third element $f(\theta', \theta'')$ is defined according to a fixed rule. He supposes the commutative and associative laws

$$f(\theta', \theta'') = f(\theta'', \theta')$$
$$f(\theta', f(\theta'', \theta''')) = f(f(\theta', \theta''), \theta''').$$

Later on, he uses instead of $f(\theta', \theta'')$ the simpler product notation $\theta' \cdot \theta''$.

Kronecker next proves the *fundamental theorem on finite abelian groups*, which asserts, for every finite abelian group, the existence of a "fundamental system" of generators $\theta_1, \theta_2, \dots$ of orders $n_1, n_2, \dots$, such thst

1) the expression

$$\theta_1^{h_1} \theta_2^{h_2} \theta_3^{h_3} \dots \qquad (h_i = 1, 2, \dots, n_i)$$

represents all elements θ, and every element just once.

2) every n_i is divisible by n_{i+1}

3) the product $n_1 n_2 \dots$ is equal to the order of the group.

Kronecker notes that this theorem is in full accordance with the results of Schering.

We now turn to England.

Arthur Cayley

In 1854, Cayley published two papers under the title "On the Theory of Groups, as Depending on the Symbolic Equation $\theta^n = 1$" in Philos. Magazine of the Royal Society London, Vol. 7. He starts with a symbol Θ representing an operation on quantities $x, y, \dots$. He writes

$$\Theta(x, y, \dots) = (x', y', \dots).$$

If $x', y', \dots$ represent a permutation of $x, y, \dots$, the operation Θ is "what is termed a substitution", says Cayley. If the operand is a single quantity x, the symbol Θ is an "ordinary function symbol"

$$\Theta x = x' = f x.$$

The symbol 1 "will naturally denote the operation which leaves the operand unaltered", and $\Theta\Phi$ denotes the "compound operation". Cayley notes that these symbols Θ are not in general commutative, but are associative.

Next, Cayley introduces the notion of a group table. He says

A set of symbols

$$1, \alpha, \beta, \ldots$$

all of them different, and such that the product of any two of them ... belongs to the set, is said to be a *group*. It follows that if the entire group is multiplied by any one of the symbols, either as further or as nearer factor, the effect is simply to reproduce the group; or what is the same thing, that if the symbols of the group are multiplied together as to form a table, thus:

Further factors

	1	α	β	...
1	1	α	β	
α	α	α^2	$\beta\alpha$	
β	β	$\alpha\beta$	β^2	

(nearer factors, left column)

that as well each line as each column of the square will contain all the symbols $1, \alpha, \beta, \ldots$.

The next step towards the abstract definition of groups was made by Cayley in 1878. In his paper "The Theory of Groups", American Journal of Mathematics 1, p. 50–52 he writes:

A set symbols $\alpha, \beta, \gamma, \ldots$ such that the product $\alpha\beta$ of each two of them (in each order, $\alpha\beta$ or $\beta\alpha$) is a symbol of the set, is a group ...

A group is defined by the laws of combination of its symbols.

and he formulates the problem: to find all finite groups of a given order n. He also states that every finite group of order n can be represented as a group of substitutions upon n letters. He say:

But although the theory as above stated is a general one, including as a particular case the theory of substitutions, yet the general problem of finding all groups of a given order n, is really identical with the apparently less general problem of finding all groups of the same order n which can be formed with the substitutions upon n letters.

To prove this proposition, Cayley uses the group table. He tacitly supposes that the multiplication of the symbols $\alpha, \beta, \gamma, \ldots$ in his definition of a group is associative:

$$(\alpha\beta)\gamma = \alpha(\beta\gamma).$$

He also supposes that the group has a unit element, and that every line and every column of the multiplication table contains all elements of the group. We may say that the abstract notion of a finite group was present in his mind, but that he did not clearly state the conditions the multiplication $\alpha\beta$ has to satisfy. In this respect, Kronecker's earlier paper of 1870 on abelian groups was better, for Kronecker clearly formulated the commutative and associative laws.

In 1882, the abstract notion "group" was defined with complete clarity, nearly simultaneously, by Walter von Dyck and Heinrich Weber. It so happens that their papers were both published in one and the same Volume 20 of the Mathematische Annalen. The paper of von Dyck is dated "Leipzig, am 6.

Dezember 1881", and that of Weber is dated "Königsberg in Preussen, Mai 1882". The wording of their definitions is completely different, as we shall see presently.

Walter von Dyck

Walter Dyck, as he calls himself in Math. Annalen 20, was Felix Klein's assistent at Leipzig from 1871 to 1884. He was strongly influenced not only by Klein, but also by Hamilton and Cayley. At the top of his paper "Gruppentheoretische Studien" in Math. Annalen 20, p. 1–44, he inserts a quotation from Cayley:

A group is defined by means of the laws of combination of its symbols.

Right at the beginning Dyck states his problem as follows:

To define a group of discrete operations, which are applied to a certain object, while abstracting from any special form of representation of the single objects and supposing the operations to be given only by those properties that are essential for the formation of the group.

Dyck starts with m generating operations $A_1, \ldots, A_m$, which can be applied to an object denoted by 1. He supposes the operations to be invertible. He now applies the operations

$$A_1, A_1^{-1}, A_2, A_2^{-1}, \ldots, A_m, A_m^{-1}$$

to the object 1, and he forms composite products, which are to be read from left to right. Thus he obtains a group G.

Next he states that isomorphic groups are considered as one and the same group. Thus, what matters in the definition of a group is only the law of composition of its elements.

Operations, multiplied from left to right, automatically obey the associative law, so Dyck's groups satisfy the modern "group axioms". Conversely, if one starts with an abstract group in the modern sense, one can always interpret the group elements A as operators, acting on the group by multiplying all group elements by A on the right. So Dyck's definition of a "discrete group" is equivalent to the modern definition of a finitely generated abstract group.

Dyck next notes that the structure of the group is known as soon as one knows the relations between its generators, which can be written as

$$F_h(A_1, \ldots, A_m) = 1.$$

He first considers the case of a group G without relations, a "free group" as we call it today. In this case every element of the group can be written in only one way as a produt of factors A_i and A_i^{-1}. In order to avoid negative exponents, Dycks introduces a new generator A_n defined by the relation

(1) $A_1 A_2 \ldots A_m A_n = 1.$

Now, if G is such a free group, and if $\bar{G}$ is any group generated by $\bar{A}_1, \ldots, \bar{A}_m$ in which certain relations $F_h = 1$ hold, then the expressions F_h generate a normal subgroup H in G, and there is a morphism $G \to \bar{G}$ $(A_i \to \bar{A}_i)$ with kernel H. In modern terminology, we have an isomorphism

(2) $$\bar{G} \cong G/H.$$

Dyck also shows that the free group G can be represented as a group of fractional linear transformations

(3) $$z' = \frac{az+b}{cz+d}$$

as follows. One starts with a domain bounded by $m+1$ arcs of circles inside a fixed circle and perpendicular to it (the shaded domain in Fig. 26). Let the vertices, in which two of the bounding arcs come together, be $a_1, \ldots, a_m, a_n$. The inversion with respect to the arc $a_1 a_n$ (or the reflection if $a_1 a_n$ is a line segment) transforms the "shaded domain" onto a "white domain". The operation A_i is defined to be the product of the inversions with respect to the two adjacent circles $a_{i-1} a_i$ and $a_i a_{i+1}$. The relation (1) is satisfied, and the operators $A_1, \ldots, A_m$ generate a free group G. The shaded and the white domain form together a fundamental domain of the group G.

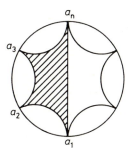

Fig. 26. Dyck's construction of a fundamental domain

Heinrich Weber

Weber's paper "Beweis des Satzes, dass jede eigentlich primitive quadratische Form unendlich viele Primzahlen darzustellen fähig ist" was composed in May 1882 and published in Math. Annalen 20, p. 301–329. The first section is entitled "Hilfssätze über Gruppen". It starts with the following definition:

A system G of elements of any kind, $\Theta_1, \Theta_2, \ldots, \Theta_h$ is called a *group of order h*, if it satisfies the following conditions:

1. From any two elements of the system one derives a new element of the same system by a prescription, which is called composition or *multiplication*. In signs

$$\Theta_r \Theta_s = \Theta_t.$$

2. One always has

$$(\Theta_r \Theta_s) \Theta_t = \Theta_r (\Theta_s \Theta_t) = \Theta_r \Theta_s \Theta_t.$$

3. From $\Theta \Theta_r = \Theta \Theta_s$ or from $\Theta_r \Theta = \Theta_s \Theta$ it follows that $\Theta_r = \Theta_s$.

From these postulates Weber derives, always for finite groups, the existence of a unity Θ_0 and of an inverse element Θ^{-1} to every element Θ.

The restriction to finite groups was dropped by Weber in 1893. In his paper "Die allgemeinen Grundlagen der Galois'schen Gleichungstheorie", Math. Annalen 43, p. 521–549 Weber defines:

A system S of things (elements) of any kind in finite or infinite number becomes a *group*, if the following assumptions are fulfilled:

1) A prescription is given, according to which from any first and second element of the system a definite third element of the same system is derived.

Weber next introduces the notation AB, and he says that the commutative law is not presupposed, but:

2) the associative law is assumed ...
3) it is supposed that, if $AB = AB'$ or $AB = A'B$, then necessarily $B = B'$ or $A = A'$ must hold.

For finite groups one obtains, as a consequence of 1), 2), 3):

4) If two of the three elements A, B, C are taken arbitrarily in S, the third can always be determined in such a way that
$$AB = C$$
holds.

Weber proves 4) for finite groups, and he continues:

For infinite groups this proof is not conclusive. For infinite groups we will include the property 4) as a postulate in the definition of the notion group.

The same definition was also given in Weber's very influential "Lehrbuch der Algebra", right at the beginning of Volume 2 (dated Strassburg, July 1886).

Later investigations concerning the reduction of the group axioms and their dependence are outside the scope of the present chapter.

Part D
The Structure of Finite Groups

After the introduction of the notion Abstract Group by Cayley, Kronecker, Dyck, and Weber, the theory of groups changed its character. Formerly the main problems were: to determine the structure of permutation groups under certain conditions such as transitivity and primitivity, and to determine the structure of finite-dimensional continuous groups of transformations. Afterwards, the problem was: to find general theorems concerning the structure of abstract groups, to determine all finite groups of a given order h, and so on.

Some fundamental structure theorems for finite groups have been discussed already in Parts A and C of the present chapter, namely:

1. The "Main Theorem" on finite abelian groups, which says (in modern terminology) that every finite abelian group is a direct product of cyclic groups. The orders $n_1, n_2, \ldots$ of these groups can be chosen in such a way that each n_i is divisible by n_{i+1}. Alternately, they may be chosen as powers of primes. As we have seen, this theorem was proved by Kronecker in 1870. The uniqueness of the factors n_i was proved by Frobenius and Stickelberger in 1879 (Crelle's Journal für Math. 86, p. 217–262). The generalization of these theorems to finitely generated abelian groups is easy (see my Algebra II, fifth ed., p. 8).

2. Sylow's theorems, which say: If the order h of a finite group is divisible by p^a, but not by a higher power of the prime p, there is at least one subgroup of order p^a. All these subgroups are conjugate, and their number is of the form $kp+1$.

3. Every finite group of order p^a contains at least one element of order p permutable with all group elements. The factor group with respect to the subgroup generated by this element contains again such an element of order p, and so on, so the group is solvable. This theorem too was proved in 1872 by Sylow.

The further development of the theory of finite groups at the end of the nineteenth century is to a large extent due to Otto Hölder. We shall now discuss the mathematical work of this remarkable scholar, and also his little known work in the philosophy of mathematics.

Otto Hölder

Otto Hölder (1859–1937) was an excellent mathematician. Like his great teacher Weierstrass, he was one of those who gave a new direction to modern mathematics: the direction from formal calculations and heuristic thinking to a rigorous, critical attitude. It is a pleasure to read his lucid papers.

For a short biography of Hölder I may refer to my obituary "Otto Hölder", Sitzungsber. sächs. Akademie Leipzig 90, p. 97–101 (1938). From this abituary I quote: "All those who knew Hölder well appreciated his quiet, kind nature and his noble character."

In Hölder's PhD thesis (Tübingen 1882) he solved the problem: If one calculates from a continuous mass density k the Newtonian potential V, under what condition does it satisfy the Laplace equation

$$\Delta V = -2\pi k?$$

To us modern mathematicians, such a problem may seem natural, but in Hölder's time even the existence of non-differentiable functions was a new and startling result. Before Weierstrass, most mathematicians tacitly assumed every continuous functions to have a derivative. Hölder's solution of the problem was highly original: the function k has to satisfy a "Hölder condition".

In the same careful way Hölder investigated the behaviour of the potential function in the vicinity of a surface on which a continuous mass distribution is given.

Under the influence of Weierstrass Hölder next moved into complex function theory. The well-known "Theorem of Casorati-Weierstrass" concerning the behaviour of a complex function in the neighbourhood of an isolated singularity was first correctly formulated by Casorati, but his paper, in Italian, was very difficult to understand. Next, Weierstrass proved the theorem for the case of a function which is regular in the whole z-plane, its only singularity being at infinity. Hölder was the first to state the theorem with complete clarity and full generality, and to present an extremely simple proof. See on this subject E. Neuenschwander: The Casorati-Weierstrass Theorem, Historia Math. 5, p. 139–166 (1978).

In 1884, Hölder became "Privat-Dozent" at Göttingen. To obtain the "Venia legendi", the right to lecture, he had to present a "Habilitationsschrift". In this paper, entitled "Zur Theorie der trigonometrischen Reihen" (Math. Annalen 24, p. 181–216) Hölder investigates the representation of arbitrary (possibly not bounded) functions by Fourier series. In order to define the Fourier coefficients as integrals, he had to introduce a new definition of the integral of an unbounded function.

Early in his career, Hölder had tried to obtain an algebraic differential equation for the Gammafunction, without success. In 1886, he succeeded in proving that such a differential equation is impossible.

Most important for our present purpose are Hölder's classical investigations on Galois theory and on finite groups. His first paper on these subjects was entitled "Zurückführung einer Gleichung auf eine Kette von Gleichungen", Math. Annalen 34, p. 26–56 (1889). If one wants to solve an equation by radicals, it is sometimes necessary to introduce "accessory irrationalities", that is, quantities that cannot be expressed as rational functions of the roots with coefficients from a given ground field. Hölder now asks: Is it possible to solve any given equation by adjoining the roots of a sequence of auxiliary equations such that

a) no accessory irrationalities are introduced,

b) after the adjunction of the roots of all preceding auxiliary equations the Galois group of each auxiliary equation is simple?

The solution is, of course: To any composition series of the Galois group corresponds a sequence of field extensions as required. Hölder now defines the notion "factor group", and he shows that the Galois groups of the single auxiliary equations are just the simple factor groups of the composition series.

Hölder next shows that these factor groups are uniquely defined but for their order and but for isomorphisms. This is the famous "Jordan-Hölder Theorem". As we have seen, Jordan had already proved that the indices in the composition series are unique but for their order.

The Jordan-Hölder theorem can be extended to infinite groups. If a group has a composition series, all such series have the same factor groups up to isomorphy, and every normal subgroup is a member of a composition series. These two statements are consequences of a still more general *Refinement*

Theorem due to Otto Schreier, which is also valid for groups with operators. See O. Schreier: Abhandlungen math. Seminar Hamburg 6, p. 300–302.

In a later paper, Hölder proved that an irreducible cubic equation having three real roots cannot be solved by real radicals.

In a sequence of papers published in Math. Annalen during the years 1892–1895, Hölder systematically investigated the structure of finite groups. The titles of these papers are:

Über einfache Gruppen. Math. Ann. 40, p. 55–88 (1892).

Die Gruppen der Ordnungen p^3, pq^2, pqr, p^4. Math. Ann. 43, p. 301–412 (1893).

Bildung zusammengesetzter Gruppen. Math. Ann. 46, p. 321–422 (1895).

The problem considered in this last paper is: If the structures of a factor group G/H and of the normal divisor H are given, how can one find all possible structures of the group G? Hölder, and later Otto Schreier, developed methods to solve this problem. See O. Schreier: Ueber die Erweiterung von Gruppen I, Monatshefte für Mathematik und Physik 34, and II, Abhandlungen Math. Seminar Hamburg 4, p. 321–346 (1926).

In another paper Hölder determined the structure of all groups of square-free order. Such a group G always has a cyclic center H, the factor group G/H being cyclic too. Hölder also determined all representations of such groups by linear transformations over the field of complex numbers.

During the years 1914–1923 Hölder mainly occupied himself with logical-philosophical questions. These investigations finally resulted in an excellent, though little known book entitled "Die mathematische Methode". I have tried to summarize the leading ideas of this book in my obituary. What follows is a free translation from this obituary.

According to Hölder one of the essential characteristics of the mathematical method can be described as building new notions as a superstructure to notions present at a certain stage, in the following sense. The notions and methods applied at a certain stage are envisaged as objects of the mathematical investigation at a higher stage. For instance: one applies a certain algorithm or method of proof, and afterwards one considers the scope and the limits of this method, making the method itself an object of investigation. From this, Hölder concludes that it is impossible to comprehend the whole of mathematics by means of a logical formalism, because logical considerations concerning the scope and the limits of the formalism necessarily transcend the formalism and yet belong to mathematics. This conclusion is fully confirmed by later investigations of Kurt Gödel.

Hölder also considers the question, how subsets of a set can be defined. He concludes that the notion "set of all subsets" is not admissible. If this conclusion is admitted, it follows that Dedekind's definition of the set of real numbers by means of cuts cannot be accepted. Thus, Hölder is forced to postulate the existence of the continuum of real numbers by special axioms.

I have tried to give the reader an idea of the scope of the work of this excellent mathematician and philosopher of mathematics. Now let us return to group theory.

Finite Linear Groups

Camille Jordan proved in 1878 (Crelle's Journal für Math. 84) that every finite group of linear transformations, say G, has an abelian normal subgroup H such that the index i does not exceed a bound depending only on the number of variables. The subgroup H can be diagonalized, that is, its matrices consist of blocks, each block being a multiple λI of a unit matrix I. The blocks correspond to linear subspaces of the complex n-dimensional vector space on which G operates, the whole vector space being the direct sum of the subspaces. If there are several subspaces with different values of λ, the group G is "imprimitive", that is, it permutes the subspaces. Hence, if G is primitive, the subgroup H consists of multiples λI of the unit matrix I. In this case, the factor group G/H can be regarded as a group of projective transformations, and the order of this projective group is bounded.

Explicit limits for the index i were given by Bieberbach (1911), Frobenius (1911), Blichfeldt (1916), and Speiser (1927). For more details see the excellent exposition of A. Speiser: Die Theorie der Gruppen von endlicher Ordnung (fourth edition, p. 194–202).

In 1898, E.H. Moore proved a theorem of fundamental importance (Math. Annalen 50, p. 213–214). namely:

Every finite group of linear transformations with complex coefficients leaves invariant a positive Hermitian form

(1) $$\Sigma\, a_{ik}\, \bar{x}_i\, x_k \qquad (a_{ik} = \bar{a}_{ki}).$$

Moore's proof is very simple: one applies to the unit form

$$\Sigma\, \bar{x}_k\, x_k$$

all transformations of the group, and one adds the results.

The same result has already been obtained in 1896 by A. Loewy. It was published without proof in a Comptes-Rendus note entitled "Sur les formes définies à indeterminées conjuguées de M. Hermite" (C.R. Acad. Paris 123, p. 168–171).

Moore's method of constructing an invariant positive Hermitean form can be extended to compact Lie groups. One has only to replace the summation by an integration over the Lie group, using an invariant volume element dV, according to an idea of Adolf Hurwitz. His paper "Über die Erzeugung von Invarianten durch Integration" was published in Nachrichten der Ges. der Wiss. Göttingen 1897, p. 71–90.

Even more generally: If the matrix elements of a group of complex linear transformations are bounded, the group leaves invariant a positive Hermitean form. This was proved by H. Auerbach in 1932 (Comptes Rendus Acad. Paris 195, p. 1367).

Now let us return to finite groups. As we have seen, Felix Klein has determined all finite binary projective groups, that is, all finite groups of fractional linear transformations

$$z' = \frac{a z + b}{c z + d}$$

with complex coefficients a, b, c, d. A simple direct derivation of these groups was given by H.H. Mitchell in 1911 (Transactions Amer. Math. Soc. 12, p. 208–211).

The investigation of finite ternary projective groups was started by C. Jordan (1878) and H. Valentiner (1889) and completed by H.F. Blichfeldt (Math. Annalen 63, p. 552–572, 1907).

The real projective orthogonal group in 4 dimensions $PO(4, \mathbb{R})$ is a direct product of two subgroups isomorphic to the complex binary projective unitary group $PU(2, \mathbb{C})$:

$$PO(4, \mathbb{R}) \cong PU(2, \mathbb{C}) \times PU(2, \mathbb{C}).$$

This isomorphism was used by E. Goursat in 1889 (Annales Ecole normale (3)6, p. 9–102) to determine all finite subgroups of $PO(4, \mathbb{R})$. The corresponding groups of four-dimensional rotations have been determined by W. Threlfall and H. Seifert in 1931 (Math. Annalen 104, p. 1–70).

In 1905, H.F. Blichfeldt determined all primitive finite linear groups in 4 dimensions (Math. Annalen 60, p. 204–231).

For a more complete account of the theory of finite linear groups see my "Gruppen von linearen Transformationen", Springer-Verlag 1935.

Chapter 9
Lie Groups and Lie Algebras

Part A
Lie Groups

What we today call a Lie group is called by Sophus Lie and his followers a "finite continuous group". It is a connected topological group in which the elements in a neighbourhood of any group element are uniquely determined by the values of r parameters $a_1, \ldots, a_r$, which vary in an open set of a Euclidean space. The parameters may be real or complex variables.

Lie's Theory

The fundamental ideas of Lie's theory of "finite continuous groups" are already contained in his first sequence of papers on the subject, published during the years 1874–1879 (Gesammelte Abhandlungen 5, p. 1–223). However, in these early papers, the presentation of his ideas is not at all satisfactory. For instance, Lie supposed that *all* elements of an r-dimensional group of transformations can be characterized by the values of r parameters $a_1, \ldots, a_r$. Later on Lie realized that in many cases this parametrization is valid only locally, in a neighbourhood of every group element. Also, in his definition of the notion "group", Lie only required products of group elements to be in the group, and he claimed that a transformation group in this sense must necessarily contain the identity and the inverse of every group element. Later on he recognized that the existence of the identity and of the inverse T^{-1} must be postulated.

In December 1880, Lie presented to the Mathematische Annalen a paper entitled "Theorie der Transformationsgruppen" (Math. Annalen 16, p. 441–528), in which he gave a better exposition of his theory. An English translation of this paper, with a very useful commentary, was published in 1975 by M. Ackermann and R. Hermann under the title "Sophus Lie's 1880 Transformation Group Paper" (Math. Sci. Press, Brookline, Mass.).

In earlier papers, Lie had not specified the nature of the functions defining the transformations of a group. In his Annalen paper, Lie supposed these functions to be analytic functions, defined by power series in the neighbourhood of any point of their domain of definition. The groups were supposed to contain the inverses T^{-1} of their elements, and some errors in Lie's earlier papers were corrected.

A still better exposition of the whole theory was given in a monumental three-volume work of Sophus Lie and Friedrich Engel entitled "Theorie der Transformationsgruppen" (Teubner, Leipzig, 1888–1893). This work was very influential: we all learnt the principles of Lie's theory from it. In the following exposition I shall use this standard work as my main source.

From the very beginning, Lie and Engel consider transformations

$$(1) \qquad\qquad x_i' = f_i(x_1, \ldots, x_n) \quad (i = 1, \ldots, n)$$

defined in a certain domain D by means of analytic functions f_i, whose functional determinant is different from zero. Hence, if one restricts oneself to a suitable neighbourhood of any point of D, the transformations are invertible.

Next the authors consider transformations (1) depending on r parameters $a_1, \ldots, a_r$:

$$(2) \qquad\qquad x_i' = f_i(x_1, \ldots, x_n, a_1, \ldots, a_r).$$

Again, the functions f_i are supposed to be analytic functions of the x and a, defined in a certain domain. Moreover, the authors suppose that the parameters are *essential*, that is, that it is not possible to represent the same set of transformations by less than r parameters.

If the product of two transformations (2) always belongs to the set, the set is called a *finite continuous group*. It is shown that the composition of two group elements is defined by analytical functions

$$(3) \qquad\qquad c_k = \varphi_k(a_1, \ldots, a_r, b_1, \ldots, b_r).$$

This theorem and many later theorems are valid only locally, in a neighbourhood of the identity transformation and in a neighbourhood of a point x_0 of the manifold on which the group operates. Very often, manifolds and groups can be parametrized only locally, and the product of two group elements in a neighbourhood of unity can very well fall outside this neighbourhood. Therefore, if one wants to obtain a rigorous theory, one has to restrict oneself to suitable neighbourhoods of the identity I. For instance, instead of requiring that products ST are always in the group, one may postulate: "Every neighbourhood U of the element I contains a sub-neighbourhood V of I such that, if S and T are in V, the product ST is an element of U". See O. Schreier: Abstrakte kontinuierliche Gruppen, Abhandlungen math. Seminar Univ. Hamburg 4 (1925) p. 15–32. If all enunciations of Lie and Engel are modified in this way, one obtains a rigorous theory of Lie groups, or rather of "group kernels", that is, of open sets in the space of the parameters a in which products ST and inverses S^{-1} are defined locally.

On page 22, the authors note that a finite continuous group in their sense does not necessarily contain the identity and the inverse transformation T^{-1} to every element T. In later enunciations, they often add an extra condition like "if the group contains the identity ...". In what follows, I shall always presuppose the existence of I and T^{-1}.

Infinitesimal Transformations

A fundamental idea in Lie's theory is the passage from transformations T to what Lie calls *infinitesimal transformations*. Lie obtains them, in all of his papers, by differentiating the transformations (2) with respect to the parameters a, and he writes them first in the form

(4) $$\delta x_i = X_i(x)\,\delta t.$$

Next he considers the infinitesimal transformations as linear operators on functions $F(x_1, \ldots, x_n)$. He puts

(5) $$AF = X_1 \frac{\partial F}{\partial x_1} + \ldots + X_n \frac{\partial F}{\partial x_n}.$$

This way of expressing infinitesimal transformations as differential operators allows him to form commutators

(6) $$(A, B) = AB - BA.$$

The idea to consider linear differential operators (5) is due to Jacobi, who also introduced the "Jacobi symbol" (A, B) and proved the "Jacobi identity"

$$(A, (B, C)) + (B, (C, A)) + (C, (A, B)) = 0.$$

Lie next proves: The infinitesimal transformations belonging to an r-dimensional Lie group are linear combinations of r linearly independent infinitesimal transformations:

(7) $$A = \lambda_1 A_1 + \ldots + \lambda_r A_r,$$

and if A and B are in the set (7), so is their commutator (A, B).

Conversely, if $A_1, \ldots, A_r$ are linearly independent infinitesimal operations such that the set (7) also contains the commutators (A, B), Lie states that each of the infinitesimal transformations of the set generates a one-dimensional group, and that the union of these one-dimensional groups is an r-dimensional continuous group. This theorem is locally true, and in fact, Lie's proof is valid only locally.

Today, linear sets (7) in which products (A, B) are defined satisfying the postulates

$$(B, A) = -(A, B)$$

$$(A, (B, C)) + (B, (C, A)) + (C, (A, B)) = 0$$

are called "Lie algebras". As a first introduction to the theory of Lie groups and Lie algebras I can recommend:

P.M. Cohn: Lie groups. Cambridge Tracts in Mathematics 46 (1957).

The fact that the commutators (A, B) are in the linear set (7) implies

(8) $$(A_i, A_k) = \Sigma\, c_{iks} A_s.$$

The constants c_{iks} depend only on the rules of composition of the group elements, not on the particular representation of the group as a group of transformations. Isomorphic Lie groups have, for a suitable choice of the basis elements of the linear set (7), the same constants c_{iks}. Even more: if two groups G and H are *locally isomorphic*, their constants c_{iks} are the same, or in other words: *Locally isomorphic groups define isomorphic Lie algebras.*

In the book of Lie and Engel the notion "locally isomorphic" does not occur. On page 291 of Vol. I the authors assert: "If two r-dimensional Lie groups have the same constants c_{iks}, they are isomorphic". This is not true. For instance, the group of rotations about the origin in the complex z-plane

$$z' = e^{ia} z$$

is locally isomorphic, but not isomorphic to the group of translations of the real line

$$x' = x + a.$$

It is very curious that Lie and Engel did not note this. They solved the differential equations determining the composition of the group by means of power series, which converge in a neighbourhood of the unity element, and they left it at that.

Three Fundamental Theorems

In Chapter 25 of Volume 3 of their "Theorie der Transformationsgruppen", Lie and Engel summarized the essentials of the theory in three fundamental theorems. In order to retain some of the flavour of the text, I shall present a slightly abridged, but fairly literal translation.

First Fundamental Theorem. If a set of ∞^r transformations

$$(9) \qquad x_i' = f_i(x_1, \ldots, x_n; a_1, \ldots, a_r)$$

is an r-dimensional group, the x_i', considered as functions of the a, satisfy differential equations of the form

$$(10) \qquad \frac{\partial x_i'}{\partial a_k} = \sum_{j=1}^{n} \Psi_{jk}(a_1, \ldots, a_r) \, \xi_{ji}(x_1', \ldots, x_n'),$$

in which the determinant of the Ψ_{jk} is not identically zero, the ξ_{ji} being such that the r expressions

$$X_k f = \sum_{i=1}^{n} \xi_{ki}(x_1, \ldots, x_n) \frac{\partial f}{\partial x_i}$$

represent independent infinitesimal transformations.

Conversely, if a set (9) of ∞^r transformations satisfies differential equations of the form (10) and contains the identical transformation, while the determinant of the $\Psi_{jk}(a)$ has a finite value different from zero for the identical transformation, then the set (9) is an r-dimensional group with pairwise inverse transformations. It coincides with the union of all one-dimensional groups generated by the ∞^{r-1} infinitesimal transformations.

$$XF = \sum_1^r \lambda_k X_k F.$$

Second Fundamental Theorem. Any r-dimensional group of transformations (9) consisting of pairwise inverse transformations contains r independent infinitesimal transformations

$$X_k f = \sum_{v=1}^n \xi_{kv}(x_1, \ldots, x_n) \frac{\partial f}{\partial x_v}$$

satisfying relations of the form

(11) $$(X_i, X_k) f = \sum_{s=1}^r c_{iks} X_s f,$$

and the group is the union of the ∞^{r-1} one-dimensional subgroups generated by the infinitesimal transformations $\sum \lambda_k X_k f$.

Conversely r independent infinitesimal transformations $X_1 f, \ldots, X_r f$ satisfying relations of the form (11) always generate an r-dimensional group with pairwise inverse transformations.

Third Fundamental Theorem. If

$$X_k f = \sum_{v=1}^n \xi_{kv}(x_1, \ldots, x_n) \frac{\partial f}{\partial x_v}$$

are independent infinitesimal transformations of an r-dimensional group, which implies that they satisfy relations of the form (11), then the r^3 constants c_{iks}, which determine the composition of the group elements, satisfy the equations

$$c_{iks} + c_{kis} = 0$$

(12) $$\sum_{\tau=1}^r (c_{ik\tau} c_{\tau js} + c_{kj\tau} c_{\tau is} + c_{ji\tau} c_{\tau ks}) = 0.$$

Conversely, if one knows r^3 constants satisfying the relations (12) and if n is sufficiently large, there exist r independent infinitesimal transformations X_i which satisfy (11) and hence generate a group with just these constants c_{iks}.

Every one of these theorems consists of two halves. The first half is valid for every Lie group G, and also for every locally Euclidean "group kernel" in which products are defined in a neighbourhood of the unity element. The second halves, the converse theorems, are valid only locally. To illustrate this, let us consider an example.

Consider the upper half of the unit circle in the (x, y)-plane. Projecting it on its diameter, we can characterize every point by its first coordinate x. Now consider an infinitesimal rotation of the circle about its centre:

$$\delta x = -\sqrt{1-x^2}\,\delta t.$$

Integrating this infinitesimal rotation, one obtains a "group kernel" consisting of rotations with angles between $-180°$ and $+180°$. Rotations with larger angles cannot be represented by transformations of the line segment between -1 and $+1$ on the x-axis.

It follows that the first and the second fundamental theorem are valid only locally, in a neighbourhood of the identity transformation.

For the third theorem the situation is more complicated. If a Lie algebra is given, that is, if the coefficients c_{iks} are known, it can be proved that a *Lie group kernel* exists having just these constants. This was proved by F. Schur in 1889. However, the existence of a *global* Lie group corresponding to a given Lie algebra was proved only much later by Elie Cartan.

For a thorough discussion of the whole problem of the rigorous foundation of Lie's fundamental theorems I may refer the reader to:

F. Schur: Neue Begründung der Theorie der endlichen Transformationsgruppen, Math. Annalen 35, p. 161–197 (1890);

E. Cartan: La théorie des groupes finis et continus et l'Analysis Situs, Paris 1930;

E. Cartan: La théorie des groupes finis et continus et la géométrie differentielle, Paris 1937;

D. Montgomery and L. Zippin: Topological Transformation Groups, New York 1955.

In the last mentioned paper the authors show that it is not necessary to suppose, as F. Schur had done, that the composition functions (3) have second continuous derivatives. If the functions are continuous, one can make them analytical by introducing new variables.

For more information about the history of Lie group theory see H. Freudenthal: L'algèbre topologique, en particulier les groupes topologiques et de Lie, Revue de Synthèse (3), Nos 49–52, p. 223–243 (1968).

Part B
Lie Algebras

The main results of Part A and their immediate consequences may be summarized thus:

Every Lie group defines a Lie algebra. Conversely, if the Lie algebra is given, the structure of the Lie group is completely determined locally, that is, two Lie groups having isomorphic Lie algebras are locally isomorphic. The local Lie subgroups of the Lie group are determined by the subalgebras of the Lie algebra. If the Lie group is locally simple, that is, if it has no locally defined invariant Lie subgroup, the Lie algebra is simple, that is, it has no ideal except itself and the zero ideal. Hence, to investigate the structure of Lie groups and in particular of simple Lie groups, one has to investigate the structure of Lie algebras and in particular of simple Lie algebras.

The structure theory of Lie algebras is mainly due to Wilhelm Killing and Élie Cartan. I shall now sketch the history of this structure theory.

Sophus Lie and Friedrich Engel

In 1883, Lie determined all simple Lie algebras of dimension r having maximal subalgebras of dimensions $r-1$, $r-2$, and $r-3$. Later on, using a paper of M. Page (American Journal of Math. 10, 1888) Lie also succeeded in treating the case $r-4$.

In 1885, Lie published a memoir entitled "Allgemeine Untersuchungen über Differentialgleichungen, die eine continuierliche, endliche Gruppe gestatten" (Math. Annalen 25, p. 71–151. In this memoir and in a later paper (Sitzungsberichte sächs. Ges. der Wiss. 1889, p. 276–289) Lie indicated four types of locally simple Lie groups, namely

Type A: the projective linear groups $PGL(n, \mathbb{C})$ with $n > 1$,

Type B: the projective orthogonal groups $PO(2n, \mathbb{C})$ with $2n > 4$,

Type C: the projective symplectic groups $PSp(2n, \mathbb{C})$ which transform an alternating bilinear form

$$(x_1 y_2 - x_2 y_1) + \ldots + (x_{2n-1} y_{2n} - x_{2n} y_{2n-1})$$

into itself,

Type D: the projective orthogonal groups $PO(2n-1, \mathbb{C})$ with $2n-1 > 1$.

A Lie algebra is called *integrable* or *solvable* if it has a composition series

$$A \supset A_1 \supset \ldots \supset A_m = \{0\}$$

in which each composition factor A_{i-1}/A_i is a one-dimensional Lie algebra.

In 1887, Friedrich Engel published a note in Sitzungsberichte der sächsischen Ges. der Wiss. 1887, p. 89–99, in which he proved that every non-integrable Lie algebra contains a three-dimensional simple subalgebra, and conversely.

Wilhelm Killing

Wilhelm Killing (1847–1923) was a teacher of mathematics at the Lyceum Hosianum in Braunsberg (now Braniewo, Poland). The starting point of his investigations on Lie groups was the so-called "space problem". It may be stated thus: What kinds of metrical geometry are conceivable in which rigid bodies can move freely? Well-known examples are the Euclidean and Non-Euclidean geometries. The problem is: Are there other possibilities, if certain conditions are imposed on the free motions?

Killing noted that the free motions must necessarily form a Lie group. Thus, he was led to the examination of the possible structures of Lie groups.

During the years 1888–1890, Killing published a series of pioneer papers entitled "Die Zusammensetzung der stetigen endlichen Transformationsgruppen" in Math. Annalen 31, 33, 34, and 36. The importance of these papers can hardly be overestimated. Élie Cartan writes in his introduction to Vol. 1 of his "Oeuvres complètes" on page 25:

> Dans une série de mémoires parus dans les *Mathematische Annalen* de 1888 à 1890, Killing fait faire *un pas énorme* à la théorie.

And on page 26:

> Le but de mes premiers travaux, contenus dans ma Thèse (1894) a été d'exposer et de démontrer d'une manière rigoureuse les résultats obtenus par Killing ... Je n'avais donc en somme à faire qu'un travail de mise au point; je me suis attaché à mettre de l'ordre et de la précision un peu partout, à combler les lacunes qui pouvaient exister dans les démonstrations et à établir sur des bases solides celles qui reposaient sur des théorèmes inexacts.

The main results obtained by Killing are formulated by Élie Cartan in the introduction to his Thèse thus:

1. Besides the four great classes of simple groups found by Lie, there are just five possible structures of simple groups, having 14, 52, 78, 133, and 248 parameters respectively.

2. Every non-integrable Lie group is (locally) composed of an integrable invariant subgroup and another subgroup which is a direct product of simple Lie groups.

Note: An *invariant subgroup* H of a group G, characterized by the condition

$$aHa^{-1} = H \quad \text{for all } a \text{ in } G,$$

is what we today call a *normal subgroup*. The Lie algebra corresponding to such a subgroup is an *ideal* in the Lie algebra of G.

Of the two theorems 1. and 2. announced by Killing, the first has been completely proved by Cartan, as we shall see in the next section. Regarding the second theorem, Cartan notes on page 115 of his Thèse that Killing's proof is "manifestly insufficient". As we shall see, Cartan replaced 2. by a weaker theorem: Every Lie group G contains a maximal integrable invariant subgroup Γ such that G/Γ is a direct product of simple groups.

In what follows, I shall often use the terminology of Lie, Killing, and Cartan. That is, I shall speak of Lie groups and their subgroups, suppressing the word local. However, the reader should realize that all proofs presented by these ancient authors are based on the consideration of the *infinitesimal* transformations of the groups, that is, on the consideration of Lie algebras. The ground field is always $\mathbb{C}$ or (in later papers of Cartan) $\mathbb{R}$. Still later, Cartan also considered the global properties of Lie groups, but this does not belong to the subject matter of the present chapter.

Élie Cartan

Élie Cartan (1869–1951) was one of the greatest and most original mathematicians of his time. In 1894 he published his famous Thèse "Sur la structure des groupes de transformations finis et continus" (Oeuvres complètes I, p. 137–285), which is one of the most important mathematical papers ever produced.

In the introduction to his Thèse, Cartan mentions a doctoral thesis of Arthur Umlauf, entitled "Über die Zusammensetzung der endlichen continuierlichen Transformationsgruppen" (Leipzig 1891) and written under the direction of Friedrich Engel. In this thesis, some of the theorems announced by Killing are rigorously proved. The remaining gaps in Killing's demonstrations were filled by Cartan.

To explain Killing's method and Cartan's proofs, I must first say a few words on the notion *adjoint group*.

Every element t of a Lie group G induces an inner automorphism of G:

$$x \to x' = t x t^{-1}.$$

This automorphism induces an automorphism of the Lie algebra L_G:

$$X \to X' = TX.$$

The linear transformations T thus defined from a group: the *adjoint group* of the given group. To every element t of G corresponds a transformation T, and the mapping $t \to T$ is a homomorphism. The kernel of this homomorphism is the centre of G.

This homomorphism induces a homomorphism of the Lie algebra L_G onto the Lie algebra L_A of the adjoint group. To every element

$$(1) \qquad\qquad X = \sum_1^r X_i \lambda_i$$

of the Lie algebra L_G corresponds a linear transformation of the Lie algebra L_A:

$$(2) \qquad\qquad A = \sum_1^r A_i \lambda_i$$

namely

$$(3) \qquad\qquad AY = (X, Y).$$

The matrix of this linear transformation A will also be denoted by A.

The Characteristic Roots

The characteristic equation of a matrix A is

$$(4) \qquad\qquad \Delta(\omega) = \det(A - \omega I) = 0.$$

In our case, the matrix elements a_{ij} of A are linear functions of the λ_i, so $\Delta(\omega)$ is a homogeneous form of degree r in $\lambda_1, \ldots, \lambda_r$, and ω. Cartan sets

(5) $$(-1)^r \Delta(\omega) = \omega^r - \Psi_1(\lambda)\, \omega^{r-1} + \ldots + (-1)^k\, \Psi_{r-k}(\lambda)\, \omega^k$$

and he supposes that $\Psi_{r-k}(\lambda)$ is not identically zero. Now he chooses the first basis element X_1 as a "general" element such that

$$\Psi_{r-k}(\lambda) \neq 0 \quad \text{for } X = X_1.$$

The equation (4) for $X = X_1$ certainly has a root $\omega = 0$, because of

$$(X_1, X_1) = 0,$$

so k is never zero. The root $\omega = 0$ has multiplicity k for $X = X_1$.

It is well known that for every root ω_1 of the characteristic equation an eigenvector Y_1 exists such that

$$(X_1, Y_1) = \omega_1\, Y_1.$$

If the root has multiplicity $m = m_1 > 1$, one can find another eigenvector Y_2 modulo Y_1, such that

$$(X_1, Y_2) = \omega_1\, Y_2 + c_{21}\, Y_1,$$

and so on. Thus, one obtains a set of m vectors $Y_1, Y_2, \ldots, Y_m$. These elements and their linear combinations

$$Y = Y_1\, \gamma_1 + \ldots + Y_m\, \gamma_m$$

are said to *belong to the root* ω_1. They are characterized by the property

$$(A - \omega_1 I)^m\, Y = 0.$$

Altogether, there are

$$r = k + m_1 + \ldots + m_p$$

linearly independent elements Y_i belonging to the roots $0, \omega_1, \ldots, \omega_p$ of the equation (4), and these elements can be chosen as basis elements of the Lie algebra. All this is quite clear, if one considers the Jordan normal form of the linear transformation A.

If the Lie algebra L_G is solvable, it is easy to see that all roots of the characteristic equation are zero, hence we have in this case

$$(-1)^r \Delta(\omega) = \omega^r.$$

Conversely, if all roots are zero, L_G is solvable according to Engel. His proof, a little modified, is reproduced on page 46 of Cartan's thesis (Oeuvres complètes I, p. 176).

The following theorem is due to Killing:

The Lie product of two elements belonging to the roots ω_α and ω_β is zero if $\omega_\alpha + \omega_\beta$ is not a root, and belongs to ω_γ if $\omega_\alpha + \omega_\beta = \omega_\gamma$.

A special case is: If X and Y belong to the root zero, so does (X, Y). Hence the elements X belonging to the root zero form a subalgebra L_H of the Lie algebra L_G. According to the theorem of Engel L_H is solvable. The solvable Lie algebra L_H plays a fundamental role in the investigations of Killing and Cartan.

Let

$$X_\lambda = X_1 \lambda_1 + \ldots + X_k \lambda_k$$

be a generic element of the subalgebra L_H. That is: let $\lambda_1, \ldots, \lambda_k$ be indeterminates. We have formed the characteristic polynomial (5) for $X = X_1$, but we can just as well form it for $X = X_\lambda$. Cartan now proves (Thèse, p. 38 = Oeuvres complètes I, p. 168) that the roots ω_α of this polynomial are *linear functions* of the indeterminates $\lambda_1, \ldots, \lambda_k$. This theorem had been proved by Killing only for the case that H is abelian.

Following Weyl, I shall simplify the notation and write α instead of ω_α. So every root α is a linear function of $\lambda_1, \ldots, \lambda_k$.

Semi-Simple Lie Groups

A Lie group G is called *semi-simple* if it does not contain any solvable invariant Lie subgroup. For the Lie algebra L_G this means that it does not contain any solvable ideal.

In his Thèse, p. 53 (Oeuvres complètes I, p. 183) Cartan proves:

Theorem IV. Every semi-simple Lie group is a direct product of simple invariant subgroups.

So the study of semi-simple Lie groups or Lie algebras can be reduced to the study of simple Lie groups and Lie algebras.

Next, Cartan proves:

Theorem V. If G is semi-simple, $X_1, \ldots, X_k$ generate a maximal abelian subgroup H. The non-zero characteristic roots α are all simple and can be divided into pairs with zero sum: $\alpha + \alpha' = 0$, hence $\alpha' = -\alpha$. If X_α and $X_{-\alpha}$ belong to such a pair, the Lie product

$$(X_\alpha, X_{-\alpha})$$

belongs to L_H and is never zero. If α is a root, its multiples $2\alpha, 3\alpha, \ldots$ are not roots.

Because of this theorem, the defining relations of a semi-simple Lie algebra L_G can be simplified very much. As basis elements one can choose:

k elements $X_1, \ldots, X_k$ generating the subalgebra L_H,

and $r - k$ elements Y_α belonging to the $r - k$ simple roots α.

The defining relations are now

$$(X_\lambda, X_\mu) = 0$$
$$(X_\lambda, Y_\alpha) = \alpha Y_\alpha$$
$$(Y_\alpha, Y_\beta) = 0, \qquad \text{if } \alpha + \beta \text{ is not a root}$$
$$(Y_\alpha, Y_\beta) = Y_{\alpha+\beta} c_{\alpha,\beta}, \qquad \text{if } \alpha + \beta \neq 0 \text{ is a root}$$
$$(Y_\alpha, Y_{-\alpha}) = X_1 b_{\alpha 1} + \ldots + X_k b_{\alpha k}.$$

As an example, let us consider the simple group $SL(n, \mathbb{C})$, it consists of all $n \times n$-matrices with determinant 1. Its infinitesimal transformations are all matrices A having trace zero. A maximal abelian subalgebra L_H consists of all diagonal matrices

$$X_\lambda = \begin{pmatrix} \lambda_1 & & & \\ & \lambda_2 & & \\ & & \ddots & \\ & & & \lambda_n \end{pmatrix}$$

with $\lambda_1 + \lambda_2 + \ldots + \lambda_n = 0$. The roots α are the $n(n-1)$ differences $\lambda_i - \lambda_j$. The Y_α are the matrices C_{ij} having 1 in row i and column j, and zero anywhere else. We have

$$(X_\lambda, X_\mu) = 0$$
$$(X_\lambda, C_{ij}) = (\lambda_i - \lambda_j) C_{ij}$$
$$(C_{ij}, C_{jk}) = C_{ik} \qquad \text{if } i \neq k$$
$$(C_{ji}, C_{kj}) = -C_{ki} \qquad \text{if } i \neq k$$
$$(C_{ij}, C_{ji}) = X_\lambda (\lambda_i = 1, \lambda_j = -1).$$

By an elaborate investigation of all possible cases, Cartan has been able to determine all types of simple Lie groups. He finds that there are not other types than the four sequences discovered by Lie and the five exceptional groups discovered by Killing.

Cartan's derivations were simplified by Weyl, by myself, by Dynkin, and by Freudenthal. See:

H. Weyl: Theorie der Darstellung kontinuierlicher halbeinfacher Gruppen durch lineare Transformationen, Kapitel III: "Struktur der halb-einfachen Gruppen", Selecta Hermann Weyl, p. 325–347, or Math. Zeitschrift 24, p. 354–376 (1926).

B.L. van der Waerden: Die Klassifikation der einfachen Lieschen Gruppen, Math. Zeitschrift 37, p. 446–462 (1933).

E. Dynkin: Classification of Simple Lie Groups (Russian with English summary), Mat. Sbornik N.S. 18 (60), p. 347–352 (1946).

H. Freudenthal: Zur Klassifikation der einfachen Lie-Gruppen, Proceedings Akad. Amsterdam A 60, p. 379–383 (1958).

In the third part of his Thèse (Oeuvres complètes I, p. 227–286) Cartan examines the structure of non-solvable Lie groups. He shows that all solvable invariant subgroups of such a group G are contained in a maximal solvable invariant subgroup Γ, the factor group G/Γ being semisimple. In modern terminology, what he shows is that all nilpotent ideals in a Lie algebra L are contained in a maximal nilpotent ideal R, which we today call the radical of L. The residue ring L/R is a direct sum of simple Lie algebras.

Weyl's Group (S)

Let L_G be the Lie algebra of a simple Lie group G, and let L_H be, as before, the Lie algebra belonging to a maximal abelian subgroup H. The elements of L_H will now be written as

$$X_\lambda = X_1 \lambda^1 + \ldots + X_k \lambda^k.$$

Hermann Weyl proves, in §4 of his Chapter III quoted before (Selecta Hermann Weyl, p. 338–342) that the basis $(X_1, \ldots, X_k)$ of L_H can be chosen in such a way that the roots

$$\alpha(\lambda) = \alpha_1 \lambda^1 + \ldots + \alpha_k \lambda^k$$

have rational coefficients. Following Killing and Cartan, Weyl introduces the quadratic form

$$Q(\lambda) = \Sigma \, \alpha(\lambda)^2$$

and he proves that this form is positive. Using the terminology of general relativity, I shall write

$$Q = g_{ij} \lambda^i \lambda^j$$

with tacit summation. To every vector (λ^i) we may define a covector (λ_i) by the formula

$$\lambda_i = g_{ij} \lambda^j.$$

Its inversion can be written as

$$\lambda^j = g^{ji} \lambda_i.$$

The scalar product of a covector ρ and a vector λ is defined as

$$(\rho \, \lambda) = \rho_i \lambda^i$$

and the scalar product of two covectors as

$$(\rho \, \sigma) = g^{ij} \rho_i \sigma_j.$$

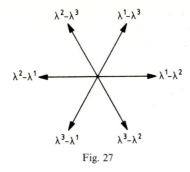

Fig. 27

Every root α defines a plane $(\alpha\lambda)=0$, and the reflection with respect to this plane is defined by

$$\rho'=\rho-2\frac{(\alpha\rho)}{(\alpha\alpha)}\cdot\alpha.$$

These reflections S_α have been considered already by Killing. They generate a finite group (S), which transforms the finite set of roots α into itself. In modern papers, this group is called the "Group of Weyl".

Let me illustrate these notions by an example. Let G be the special linear group $SL(3,\mathbb{C})$. The Lie algebra L_G consists of the 3×3-matrices with trace zero. The subalgebra L_H is formed by the diagonal matrices

$$\begin{pmatrix} \lambda^1 & & \\ & \lambda^2 & \\ & & \lambda^3 \end{pmatrix}$$

with $\lambda^1+\lambda^2+\lambda^3=0$. The non-zero roots are the 6 linear forms

$$\alpha=\lambda^i-\lambda^j \quad (i\neq j).$$

In the space of covectors, these roots may be visualized as six vectors pointing from the centre to the vertices of a regular hexagon (see Fig. 27). The reflection S_α transforms the vector α into $-\alpha$. The group (S) generated by these reflections is the dihedral group D_3, consisting of three rotations and three reflections.

For the general case see H.S.M. Coxeter: Discrete Groups Generated by Reflections, Annals of Math. 35, p. 588–621 (1933).

Real Simple Lie Algebras

The structure of *real* simple Lie algebras has been determined by Cartan in 1914 (Annales de l'École Normale 31, p. 263–355). Every Lie algebra over $\mathbb{R}$ can be extended to a Lie algebra over $\mathbb{C}$, and if the real algebra is simple, its complex extension is either simple or a direct sum of two complex conjugate simple Lie algebras. Thus, starting with the known types of simple complex Lie algebras, one can determine the real Lie algebras of which they are extensions.

Cartan's classification of real simple Lie algebras was derived by a more elegant method by F. Gantmakher (Mat. Sbornik 5, p. 101–146 and 217–249, 1939). Combining Gantmakher's methods with those of Dynkin, Freudenthal was able to obtain a still simpler derivation. See H. Freudenthal: Lie Groups (polycopy), Yale University 1961.

Among the real Lie algebras belonging to a given simple Lie algebra there is always one Lie algebra generating a *compact* Lie group, as Weyl has shown. See Selecta Hermann Weyl, p. 342–346. This fact is of fundamental importance in the representation theory of simple Lie groups, as we shall see in Chapter 14.

Part Three
Algebras

Chapter 10
The Discovery of Algebras

In the present chapter, the discovery of special algebras or "hypercomplex number systems" such as the ordinary complex numbers, the quaternions, the octonions, etcetera will be described. The general theory of the structure of algebras will be the subject of the next Chapter 11.

Complex Numbers

As we have seen in Chapter 2, Cardano was the first to introduce complex numbers like $5+\sqrt{-15}$ and $5-\sqrt{-15}$ as solutions of his problem: To divide 10 into two parts, the product of which is 40. However, the introduction of these mysterious numbers caused him "mental torture".

The next step was taken by Bombelli, who used the expressions

$$\text{più di meno } (=i)$$
$$\text{meno di meno } (=-i)$$

and who gave rules for calculating with complex numbers. As we have seen, he used cube roots of complex numbers in his solution of cubic equations in the "casus irreducibilis".

Albert Girard (1629) calls the numbers $a\pm\sqrt{-b}$ *solutions impossibles*. The term *imaginary numbers* was introduced by Descartes. He writes: "For every equation one can imagine as many roots (as its degree indicates), but in many cases no quantity exists which corresponds to what one imagines." See René Descartes: La géométrie, in: Discours de la méthode (1637).

After Descartes, the leading mathematicians made free use of complex numbers. For instance, Johann Bernoulli used logarithms of complex numbers for the purpose of transforming integrals (Opera omnia, Vol. 1, p. 400). Roger Cotes, Abraham de Moivre, and Leonhard Euler knew the formula

$$(\cos x + i \sin x)^n = \cos nx + i \sin nx.$$

For Cotes and de Moivre see Ivo Schneider: Der Mathematiker Abraham de Moivre, Archive for History of Exact Sc. 5, p. 234-246 (1968). For Euler see

§ 133 of his Introductio in Analysin Infinitorum, Opera omnia, series prima, Vol. 1.

Euler visualized complex numbers as points in a plane with rectangular coordinates x, y. Introducing polar coordinates r, φ, he wrote

$$x + iy = r(\cos \varphi + i \sin \varphi)$$

and he represented the roots of the equation

$$z^n = 1$$

as vertices of a regular polygon in the z-plane. He defined the exponential function e^z for complex z, and he proved

$$e^{i\varphi} = \cos \varphi + i \sin \varphi$$

(see § 138 of the Introductio just quoted).

In his § 32 of the same Introductio Euler formulates the "fundamental theorem of algebra" as follows:

Every integer function of z can be factored into real linear or quadratic factors. Although this has not be proved rigorously yet, the truth of this statement will be more and more corroborated in what follows... .

One year later, in 1749, Euler composed a paper "Recherches sur les racines imaginaires des equations" (Mém. Acad. des Sciences Berlin, Vol. 5, published 1751), in which he defines:

On nomme quantité imaginaire celle qui n'est ni plus grande que zéro, ni plus petite que zéro, ni égale à zéro; ce sera donc quelque chose d'impossible, comme par example $\sqrt{-1}$, ou en général $a + b\sqrt{-1}$.

In §7 of the same paper Euler undertakes to prove that every real polynomial in z can be factored into real linear and quadratic factors. The idea of his proof is the same as that of the second proof of Gauss, which we have discussed in Chapter 5. However, as Gauss notes, the proof of Euler presupposes the existence of the roots in some undefined way. As we would say to-day, Euler's proof is correct if the existence of the roots in some extension field of $\mathbb{R}$ is presupposed.

Although Euler knew the geometrical representation of complex numbers by points in a plane, he did not give a satisfactory definition of the notion "complex number". Clear geometrical definitions of the addition and multiplication of complex numbers, conceived as directed line segments in a plane, were given by Caspar Wessel in 1797, by Jean Robert Argand in 1806, by John Warren in 1828, and by Carl Friedrich Gauss in 1831. The expression "complex numbers" seems to be due to Gauss.

William Rowan Hamilton defined (1843) the complex numbers as pairs of real numbers subject to conventional rules of addition and multiplication. On the other hand, Augustin Cauchy interpreted (1847) the complex numbers as residue classes of polynomials in $\mathbb{R}[x]$ modulo $x^2 + 1$.

Hamilton's Discovery of Quaternions

Analogous to the complex numbers $a+bi$ are the quaternions

$$a+bi+cj+dk$$

which William Rowan Hamilton discovered on October 16, 1843. Their multiplication is defined by the rules

$$i^2=j^2=k^2=-1,$$
$$ij=k, \qquad jk=i, \qquad ki=j,$$
$$ji=-k, \quad kj=-i, \quad ik=-j.$$

How did Hamilton arrive at this rules? What was his problem and how did he find the solution? We are well informed about these matters through documents and papers reproduced in Volume 3 of Hamilton's Mathematical Papers (Cambridge Univ. Press 1963), namely:

First, an entry in Hamilton's Note Book dated 16 October 1843 (Papers 3, p. 103–105),

second, a letter to John Graves dated 17 October 1843 (Papers 3, p. 106–110),

third, a paper in the Proceedings of the Royal Irish Academy entitled "On a New Species of Imaginary Quantities Connected with the Theory of Quaternions" and presented on November 13, 1843 (Papers 3, p. 111–116),

fourth, the Preface to Hamilton's "Lectures on Quaternions", dated June 1853 (Papers 3, p. 117–155, in particular p. 142–144),

fifth, a letter to his son Archibald which Hamilton wrote shortly before his death in 1865 (Papers 3, p. xv–xvi).

In these documents we can follow exactly each of Hamilton's steps. In this exceptional case we can observe what may go on in the mind of a mathematician when he poses himself a problem, when he approaches the solution step by step, and when, at the end, through a sort of lightning stroke, he so modifies the problem that it becomes solvable.

Hamilton knew and used the geometrical representation of complex numbers, but in his published papers he preferred the definition of complex numbers as pairs of real numbers (a, b). He now posed himself the problem: To find out how number-triplets (a, b, c) are to be multiplied in analogy to pairs (a, b).

In the letter to his son just mentioned, Hamilton writes:

Every morning in the early part of the above-cited month (October 1843), on my coming down to breakfast, your brother William Edward and yourself used to ask me: "Well, Papa, can you multiply triplets?" Whereto I was always obliged to reply, with a sad shake of the head, "No, I can only add and subtract them".

In analogy to the complex numbers $a+bi$, Hamilton wrote his triplets as

$$a+bi+cj.$$

He visualized his basic units $1, i, j$ as mutually perpendicular "directed segments" of unit length in space. Later on Hamilton himself used the word *vector*, which I shall also use. He sought to represented products such as

$$(a+bi+cj)(x+yi+zj)$$

as vectors in the same space. He required

first, that it be possible to multiply term by term,

secondly, that the length of the product vector be equal to the product of the lengths of the factors. He called this rule "law of the moduli".

Today we know that the two requirements can be fulfilled only in spaces of dimensions 1, 2, 4, and 8. This was proved by Adolf Hurwitz, as we shall see presently. Therefore Hamilton's attempt in three dimensions was bound to fail. His profound idea was, to pass to four dimensions.

About his first attempts we learn more from the documents. To fulfill the "law of moduli" at least for the numbers $a+bi$, Hamilton set $ii = -1$, as for ordinary complex numbers, and just so he set $jj = -1$. But what was ij and what was ji?

At first, Hamilton assumed $ij=ji$, and calculated:

$$(a+ib+jc)(x+iy+jz)=(ax-by-cz)+i(ay+bx)+j(az+cx)+ij(bz+cy).$$

Now, he asked, what is one to do with ij? Shall it have the form $\alpha+\beta i+\gamma j$?

First attempt. In Hamilton's letter to John Graves he writes:

"Its square (namely the square of ij) would seem to be $=1$, because $i^2=j^2 = -1$, and this might tempt us to take $ij=1$, or $ij=-1$; but with neither assumption shall we have the sum of the squares of the coefficients in the product $=$ to the product of the corresponding sums of squares in the factors."

Second attempt. Hamilton considered the simplest case

$$(a+ib+jc)^2=a^2-b^2-c^2+2iab+2jac+2ijbc.$$

He calculated the sum of the squares of the coefficients of 1, i, and j on the right-hand side and found

$$(a^2-b^2-c^2)^2+(2ab)^2+(2ac)^2=(a^2+b^2+c^2)^2.$$

Therefore, he said, the condition respecting the *moduli* is fulfilled, if we suppress the term involving ij altogether, and, what is more, $a^2-b^2-c^2$, $2ab$, $2ac$ are precisely the coordinates of the *square-point*, deduced in space, by a slight extension of Mr. Warren's rule for points in the plane.... In fact, if we double, in its own plane, the rotation from the positive semiaxis of x to the radius vector of the point a, b, c, we attain the direction of the radius vector drawn to $a^2-b^2-c^2, 2ab, 2ac$."

Third attempt. Hamilton reports that the assumption $ij=0$, which he made in the second attempt, subsequently did not appear to be quite right to him. He writes in the letter to Graves:

Behold me therefore tempted for a moment to fancy that $ij=0$. But this seemed odd and uncomfortable, and I perceived that the same supression of the term which was *de trop* might be attained by assuming what seemed to me less harsh, namely that $ji=-ij$. I made therefore $ij=k$, ji $=-k$, reserving to myself to inquire whether k was 0 or not.

Fourth attempt. Somewhat more generally, Hamilton multiplied $a+ib+jc$ and $x+ib+jc$. In the letter to Graves, Hamilton concludes:

The coefficient of k still vanishes, and

$$ax-b^2-c^2, \quad (a+x)b, \quad (a+x)c$$

are easily found to be the correct coordinates of the *product-point* in the sense that the rotation from the unit line to the radius vector of a, b, c being added in its own plane to the rotation from the same unit line to the radius vector of the other factor-point x, b, c, conducts to the radius vector of the lately mentioned product-point; and that the latter radius vector is in length the product of the two former. Confirmation of $ij=-ji$, but no information yet of the value of k.

The Leap into the Fourth Dimension

After this encouraging result Hamilton said to himself: "Try boldly then the general product of two triplets" (letter to Graves, Hamilton's Papers 3, p. 107). He calculated

$$(a+ib+jc)(x+iy+jz)=(ax-by-cz)+i(ay+bx)+j(az+cx)+k(bz-cy).$$

He tried to set $k=0$ and asked: Is the law of the moduli satisfied? In other words: does the identity

$$(a^2+b^2+c^2)(x^2+y^2+z^2)=(ax-by-cz)^2+(ay+bx)^2+(az+cx)^2$$

hold? The answer is: No, the first member exceeds the second by $(bz-cy)^2$. But, says Hamilton, this is just the square of the coefficient of k in the development of the product

$$(a+ib+jc)(x+iy+jz),$$

if we grant that $ij=k$, $ji=-k$, as before.

And now comes the flash of lightning giving the entire problem a new direction. In the letter to Graves, Hamilton writes:

And here there dawned on me the notion that we must admit, in some sense, a *fourth dimension* of space for the purpose of calculating with triplets...

or, transferring the paradox to algebra, we must admit a third distinct imaginary symbol k, not to be confounded with either i or j, but equal to the product of the first as multiplier, the second as multiplicand, and therefore I was led to introduce *quaternions* such as $a+ib+jc+kd$, or (a, b, c, d).

Hamilton next explains the reasons why he thought it likely that

$$ik = iij = -j, \quad kj = ijj = -i$$

and

$$ki = j, \qquad\qquad jk = i$$

and finally

$$k^2 = ijij = -iijj = -1.$$

In Hamilton's letter to his son we learn even more about the external circumstances under which the decisive flash of insight befell him. He writes:

> But on the 16th day of the month [October 1843] which happened to be a Monday and a Council day of the Royal Irish Academy – I was walking in to attend and preside, and your mother was walking with me, along the Royal Canal...; and although she talked with me now and then, yet an undercurrent of thought was going on in my mind, which gave at last a result, whereof... I felt at once the importance. An electric circuit seemed to close, and a spark flashed forth, the herald (as I foresaw immediately) of many long years to come of definitely directed thought and work...
>
> I pulled out on the spot a pocket-book, which still exists, and made an entry there and then. Nor could I resist the impulse – unphilosophical as it may have been – to cut with a knife on a stone of Brougham Bridge, as we passed it, the fundamental formula with the symbols i, j, k:
>
> $$i^2 = j^2 = k^2 = ijk = -1,$$
>
> which contains the solution of the Problem, but of course, as an inscription has long since mouldered away.

This was written many years later. In his Note Book, written on the same day on which he had cut his fundamental formula on the stone, the course of his ideas was explained in greater detail thus:

> I believe I now remember the order of my thought. The equation $ij = 0$ was recommended by the circumstance that
>
> $$(ax - y^2 - z^2)^2 + (a + x)^2 (y^2 + z^2) = (a^2 + y^2 + z^2)(x^2 + y^2 + z^2).$$
>
> I therefore tried whether it might not be true that
>
> $$(a^2 + b^2 + c^2)(x^2 + y^2 + z^2) = (ax - by - cz)^2 + (ay + bx)^2 + (az + cx)^2,$$
>
> but found that this equation required, in order to make it true, the addition of $(bz - cy)^2$ to the second member. This *forced* on me the non-neglect of ij, and *suggested* that it might be equal to k, a new imaginary.

By underscoring the words *forced* and *suggested* Hamilton emphasizes a distinction between two mental experiences. The first was a compelling logical conclusion, which came immediately out of the calculation: one cannot set ij equal to zero, for then the law of moduli would not hold. The second experience was an insight which came over him in a flash at the canal ("an electric circuit seemed to close, and a spark flashed forth"), namely the idea that $ij = k$ might be taken to be a new imaginary unit. What Hamilton presents here is a profound psychological analysis of his own thoughts.

The same pocket book, in which Hamilton noted his rules of multiplication of the units i, j, k, also contains the formulae for the coefficients of the product

$$(a+bi+cj+dk)(\alpha+\beta i+\gamma j+\delta k)$$

and an outline of the proof of the "law of moduli".

For the sake of completeness I may mention that the law of multiplication of quaternions had already been discovered, but not published, by Gauss as early as 1820. See the article of E. Study "Theorie der gemeinen und höheren komplexen Grössen", Encyclopädie der mathematischen Wissenschaften I A 4.

Octonions

The letter to John Graves in which Hamilton announced the discovery of quaternions was written on the 17th of October 1843, one day after the discovery. The seed which Hamilton sowed fell upon fertile soil, for in December 1843 Graves already had found an algebra with 8 basis elements

$$1, i, j, k, l, m, n, o,$$

the algebra of *octaves* or *octonions*. Graves defined their multiplication as follows:

$$i^2 = j^2 = k^2 = l^2 = m^2 = n^2 = o^2 = -1$$

$$i = jk = lm = on = -kj = -ml = -no$$

$$j = ki = ln = mo = -ik = -nl = -om$$

$$k = ij = lo = nm = -ji = -ol = -mn$$

$$l = mi = nj = ok = -im = -jn = -ko$$

$$m = il = oj = kn = -li = -jo = -nk$$

$$n = jl = io = mk = -lj = -oi = -km$$

$$o = ni = jm = kl = -in = -mj = -lk$$

(see Hamilton's Mathematical Papers, Vol. 3, p. 648).

In 1848 Graves published his discovery in the Transactions of the Irish Academy 21, p. 338.

In Graves' algebra of octonions, the "law of moduli" holds, but the algebra is not associative, as Hamilton pointed out.

Octonions were rediscovered by Arthur Cayley in 1845 (Collected Papers I, p. 127 and XI, p. 368–371). Because of this the octonions are also known as *Cayley Numbers*.

A modified construction of an algebra of "octonions" was presented by Claude Chevalley. In his book "The Algebraic Theory of Spinors" (Columbia Univ. Press 1954), p. 123–128, Chevalley starts with a non-singular form Q in 8 variables which can be reduced to the form

$$Q = x_1 x_2 + x_3 x_4 + x_5 x_6 + x_7 x_8$$

and he constructs an algebra of dimension 8 with a non-associative multiplication $x * y$ such that

$$Q(x * y) = Q(x) Q(y).$$

Graves made an attempt to construct a similar algebra with 16 basis elements in which the "law of moduli" holds, but his attempt did not succeed. It could not succeed, for today we know that identities of the form

(1) $$(a_1^2 + a_2^2 + \ldots + a_n^2)(b_1^2 + b_2^2 + \ldots + b_n^2) = c_1^2 + c_2^2 + \ldots + c_n^2$$

are possible only for $n = 1, 2, 4$ or 8. I shall now give a short summary of the history of these identities.

Product Formulae for Sums of Squares

Euler certainly knew the law of the moduli for complex numbers $a + bi$:

$$(a^2 + b^2)(c^2 + d^2) = (ac - bd)^2 + (ad + bc)^2.$$

In a letter of Euler to Goldbach dated May 4, 1748, edited by P.H. Fuss (Correspondence mathématique et physique I, Petersburg 1843) we find a similar formula for sums of four squares. It agrees with Hamilton's "law of moduli" for quaternions.

The formula for 8 squares, which Graves and Cayley proved by means of octonions, had already been found by C.P. Degen in 1818. See C.P. Degen: Adumbratio demonstrationis theorematis arithmeticae maxime generalis, Mémoires de l'Académie de St. Petersbourg VIII, p. 207 (1822).

Hamilton's original problem was: Can triplets (a, b, c) and (x, y, z) be so multiplied that the law of moduli holds? In other words: Is it possible so to define u, v, w as bilinear functions of a, b, c and x, y, z that the identity

(2) $$(a^2 + b^2 + c^2)(x^2 + y^2 + z^2) = u^2 + v^2 + w^2$$

holds?

The first to show the impossibility of such an identity was Adrien Marie Legendre. In his "Théorie des nombres" (1830) he noted that the numbers 3 and 21 can be expressed as sums of three squares of rational numbers:

$$3 = 1 + 1 + 1$$
$$21 = 16 + 4 + 1$$

but that $3 \times 21 = 63$ cannot be represented in this way, since 63 is an integer of the form $8n + 7$. It follows that an identity of the form (2) is impossible, at least if one restricts oneself to bilinear forms u, v, w with rational coefficients. See A.M. Legendre: Théorie des nombres, 3rd ed., p. 198.

If Hamilton had known of this remark by Legendre he would probably have given up the search to multiply triplets. Fortunately he had not read Legendre: he was an autodidact.

The question for what values of n an identity of the form (1) is possible was finally decided by Adolf Hurwitz. With the aid of matrix multiplication he proved that $n = 1, 2, 4$ and 8 are the only possibilities. See A. Hurwitz: Über die Composition der quadratischen Formen von beliebig vielen Variablen, Nachrichten Ges. der Wiss. Göttingen 1898, p. 309–316.

Geometrical Applications of Quaternions

In 1885, Cayley showed that three-dimensional as well as four-dimensional rotations can be represented by quaternions. See A. Cayley: Recherches ultérieures sur les déterminants gauches, Crelle's Journal für Math. 50, p. 312–313. See also F. Klein: Zur Nicht-Euklidischen Geometrie, Math. Annalen 37, p. 546–554 (1890), and E. Study, Math. Papers from the Chicago Congress 1894, New York 1896, p. 376.

Let me start with the four-dimensional case. Let

$$X = s + ix + jy + kz$$

be a variable quaternion and let A and B be quaternions of norm 1. The transformation

(3) $$X' = AXB^{-1}$$

is a four-dimensional rotation, and all four-dimensional rotations can thus be obtained.

If $B = A$, one obtains a transformation

(4) $$X' = AXA^{-1}$$

which leaves the "scalar part" s of the quaternion X invariant and transforms the "vector part"

$$ix + jy + kz$$

according to a three-dimensional rotation. All three-dimensional rotations can thus be obtained. Namely: since A has norm 1, it can be written as

$$A = \cos\varphi + (ip + jq + kr)\sin\varphi,$$

the vector $v = (p, q, r)$ being a unit vector:

$$p^2 + q^2 + r^2 = 1.$$

An easy calculation now shows that the transformation (4) is a rotation about the axis v over an angle 2φ.

If the quaternion A is written in the usual form

$$A = a + ib + jc + kd$$

the coefficients a, b, c, d are called the *Cayley-Klein parameters* of the rotation. They were systematically used by Klein and Sommerfeld in their classical book "Über die Theorie des Kreisels" (1897).

If quaternions are used to represent rotations, the composition of rotations is very easy. If two rotations are obtained from quaternions A_1 and A_2, their product results from the quaternion $A_1 A_2$. The composition formulae thus obtained are identical with the formulae of Olinde Rodrigues for the composition of rotations (see the section "On Groups of Motions" in Chapter 7).

The Arithmetic of Quaternions

In 1896, Adolf Hurwitz published a highly interesting paper "Über die Zahlentheorie der Quaternionen" (Nachrichten der Gesellschaft der Wissensch. Göttingen 1896, p. 313–340), in which he developed a factorization theory of "integer quaternions", and applied it to the problem of representing integers as sums of four squares. In 1919 Hurwitz elaborated his ideas with full proofs in a very nice booklet entitled "Vorlesungen über die Zahlentheorie der Quaternionen". I shall now sketch the history of the problem and present an account of the ideas of Hurwitz.

In 1621, in his edition of the Arithmetica of Diophantos, C.G. Bachet noted that apparently every integer is either a square or a sum of 2 or 3 or 4 squares. Later authors called this assertion "Bachet's theorem". Bachet verified it for all integers up to 325. If zero terms are allowed, we may say that every integer N is a sum of four squares:

(5) $$N = s^2 + x^2 + y^2 + z^2.$$

The first proof of this assertion was given by Joseph-Louis Lagrange in 1772 (Oeuvres 3, p. 189–201). One year later, Leonard Euler presented a simpler proof (Opera omnia, Pars prima, Vol. 3, p. 218–239). For more details I may refer to Leonard Eugene Dickson: History of the Theory of Numbers, Vol. 2.

In 1828, Karl Gustav Jakob Jacobi proved that the number of representations of a given integer N as a sum of four squares of (possibly negative or zero) integers is equal to

$$24S, \quad \text{if } N \text{ is even,}$$
$$8S, \quad \text{if } N \text{ is odd,}$$

where S is the sum of all odd divisors of N. See K.G.J. Jacobi: Werke, Vol. 1, p. 239–247, and Vol. 6, p. 245–251.

In his proof, Jacobi made use of thetafunctions. A purely algebraic proof was given by Hurwitz by means of his number theory of integer quaternions. I shall now explain the fundamental ideas of his theory.

Hurwitz defines: A quaternion

$$q = s + ix + jy + kz$$

with rational coefficients s, x, y, z is called *integer*, if the coefficients are either all integers or all of the form $n + 1/2$. In other words: integer quaternions are the linear combinations of

$$\rho = \tfrac{1}{2}(1 + i + j + k), i, j, k.$$

In the ring of integer quaternions, there are 24 units. This explains the factor 24 in the expression $24S$ found by Jacobi.

The norm of an integer quaternion is a sum of four squares:

(6) $N = qq' = (s + ix + jy + kz)(s - ix - jy - kz) = s^2 + x^2 + y^2 + z^2.$

If N is even, the numbers s, x, y, z in (6) are necessarily integers, but if N is odd, they may be of the form $n + 1/2$. Hurwitz now proves that the number of representations of N as a norm of an integer quaternion is always $24S$. If N is odd, the numbers s, x, y, z are integer in only $1/3$ of the cases; this explains why for odd N the number of integer representations is only $8S$.

To prove these results, Hurwitz develops a factorization theory of integer quaternions. He first shows by means of a generalization of the Euclidean algorithm that every onesided ideal in the ring of integer quaternions is a principal ideal, that is, it is generated by one single element. It follows that any two integer quaternions have a (right or left) "largest common divisor" d.

If a quaternion q is multiplied by one of the 24 units, one obtains a right or left *associated quaternion* $q\varepsilon$ or εq. To restrict the choice of the unit ε, Hurwitz now introduces the notion *primary*. A quaternion is called primary, if it is congruent to 1 or $1 + 2\rho$ modulo $2(1 + i)$.

Now let q be any quaternion having an odd norm. Hurwitz proves that among the 24 quaternions $q\varepsilon$ (or εq) associated to q there is just one primary quaternion.

An integer quaternion π is called *prime*, if $\pi = ab$ implies that a or b is a unit. Hurwitz proves: π is prime if and only if its norm $\pi\pi'$ is an ordinary prime p. And: There are just $p + 1$ primary quaternions whose norms are equal to a given odd prime p.

Next, Hurwitz proves his fundamental factorization theorem:

Let c be a primary quaternion, and

$$N(c) = pqr\ldots$$

where $p, q, r, \ldots$ are (equal or unequal) prime factors of $N(c)$. Then c can be represented in exactly one way as a product

$$c = \pi \kappa \rho \ldots$$

where $\pi, \kappa, \rho, \ldots$ are primary quaternions having norms $p, q, r, \ldots$ in just this (arbitrary, but fixed) order.

Now it is easy to determine the number of quaternions having a given norm N. The result is $24S$.

For the extension of the methods of Hurwitz to other algebras see the papers of G. Aeberli and H. Gross in Comment. Math. Helv. 33, p. 212–239 and 34, p. 198–221.

Biquaternions

Quaternions with complex coefficients are called by Hamilton *biquaternions*. See W.R. Hamilton: Lectures on Quaternions (1853), art. 669. The algebra of these biquaternions is isomorphic to a full matrix ring over the complex number field, for quaternions $a + ib + jc + kd$ with complex coefficients a, b, c, d can be represented by matrices $\begin{pmatrix} a+bi & -c+di \\ c+di & a-bi \end{pmatrix}$.

In 1873, William Kingdon Clifford published a "Preliminary Sketch of Biquaternions", Proc. London Math. Soc. 4, p. 381–395 (Mathematical Papers, p. 181–204). In this paper, Clifford introduces two different kinds of "biquaternions". Both kinds can be written as

$$q + \omega r$$

where q and r are quaternions, while ω commutes with all quaternions. In the first part of his paper Clifford supposes

$$\omega^2 = 0$$

and he uses his biquaternions to describe rigid motions in Euclidean space. In the second part (Sections III–V) he supposes

$$\omega^2 = 1$$

and he uses the second kind of biquaternions to describe non-Euclidean motions. Introducing two new units

$$\xi = 1/2(1+\omega), \quad \eta = 1/2(1-\omega)$$

he finds

$$\xi^2 = \xi, \quad \eta^2 = \eta, \quad \xi\eta = 0,$$

so one can say, in modern terminology, that Clifford's second algebra of biquaternions is a direct sum of two quaternion algebras.

The first kind of Clifford's biquaternions was used by Eduard Study to obtain a parametric representation of the Euclidean group of rigid motions. See E. Study: Von den Bewegungen und Umlegungen, Math. Annalen 39, p. 441–566 (1891).

Full Matrix Algebras

The $n \times n$-matrices form an algebra of dimension n^2. As basis elements one can take the matrices e_{ij} having an element 1 in row i and column j, and 0 elsewhere. The multiplication rules are

$$e_{ij} e_{jk} = e_{ik}$$
$$e_{ij} e_{kl} = 0 \quad \text{if } j \neq k.$$

This algebra is called the full matrix algebra of rank n over the ground field K.

For the history of the theory of matrices I may refer to C.C. MacDuffee: The Theory of Matrices, Ergebnisse der Math. II 5 (Springer-Verlag 1933, reprinted by Chelsea). Here I shall mention only those notions and theorems that are most important for the theory of algebras.

The *characteristic function* of a matrix A is defined as a polynomial in λ:

$$F(\lambda) = \det(\lambda I - A) = \lambda^n - c_1 \lambda^{n-1} + \ldots + (-1)^n c_n.$$

This polynomial remains unchanged if A is replaced by PAP^{-1}. Its second coefficient is the *trace* of the matrix:

$$c_1 = \operatorname{tr}(A) = \Sigma a_{ii}$$

and its last coefficient is the determinant

$$c_n = \det(A).$$

By a suitable extension of the ground field F, the matrix A can be reduced to its *Jordan normal form*. This normal form consists of blocks, each block B_i having one and the same element λ_i in the main diagonal and elements 1 just above the main diagonal, all other elements being zero. Obviously, the characteristic polynomial of the normal form is

$$F(\lambda) = (\lambda - \lambda_1)(\lambda - \lambda_2) \ldots (\lambda - \lambda_n).$$

Since the characteristic function of the normal form PAP^{-1} is the same as that of A, it follows that *the roots $\lambda_1, \ldots, \lambda_n$ of the characteristic function $F(\lambda)$ are just the diagonal elements of the Jordan normal form*. These roots are called the *fundamental roots* of the matrix A.

An important theorem says: The matrix A satisfies the "fundamental equation"

(7) $$F(A) = A^n - c_1 A^{n-1} + \ldots + (-1)^n c_n I = 0.$$

This theorem was first announced by A. Cayley, and verified for $n = 2$ and $n = 3$, in his pioneer paper "Memoir on the Theory of Matrices", Philos. Transactions Royal Soc. London 148, p. 17–38 (1858). It was proved by E. Laguerre in his paper "Sur le calcul des systèmes linéaires", Journal de l'Ecole polytechnique 42, p. 215–264 (1867). Other proofs were given by Frobenius (1878), Buchheim (1885), Weyr (1890), Taber (1890), Pasch (1891), Molien (1893), and again Frobenius (1896). For full references see the article of E. Study "Theorie der gemeinen und höheren komplexen Größen", Encyklopaedie der math. Wissenschaften IA 4, p. 171 (1898).

The equation (7) can be written as

$$(A^{n-1} - c_1 A^{n-2} + \ldots \pm c_{n-1}) A = \pm c_n I.$$

Consequently, if an element A of any matrix algebra is non-singular, which means that $c_n = \det(A)$ is not zero, the matrix A has an inverse A^{-1} within the algebra.

Group Algebras

In the same paper of 1854, in which Cayley introduced the notion of an abstract group (Phil. Mag. 7, p. 40–47), he also introduced what we today call the "Group Algebra" of a finite group G. The basis elements of this algebra are just the group elements $g_1, \ldots, g_n$. In the multiplication rule

$$g_j g_k = \Sigma g_i a_{ijk}$$

the coefficients are

$$a_{ijk} = \begin{cases} 1 & \text{if } g_j g_k = g_i \\ 0 & \text{otherwise.} \end{cases}$$

Every representation of the group G by linear transformations can be extended to a representation of the group algebra. Conversely, every representation of the group algebra yields a representation of the group. Therefore the study of the structure of the group algebra is of primary importance in the theory of group representations.

The first to investigate the structure of the group algebra of a finite group was Theodor Molien. In 1893 he published an important paper "Über Systeme höherer komplexer Zahlen" in Math. Annalen 41, p. 83–158. In this paper, Molien proved several fundamental theorems concerning the structure of algebras over the complex field $\mathbb{C}$ (see Chapter 11). In a later paper entitled "Eine Bemerkung zur Theorie der homogenen Substitutionsgruppen" (Sitzungsberichte der Naturforscher-Gesellschaft der Universität Dorpat 11, p. 259–274)

Molien applied his general theory to the group algebra of a finite group. He proved that this group algebra is a direct sum of full matrix algebras, and from this he concluded that every representation of the algebra (and hence of the group) is completely reducible, and that every irreducible representation is contained in the regular representation. We shall return to this subject in Chapter 13.

Grassmann's Calculus of Extensions

As far as I know the first to define explicitly the notion of "n-dimensional vector space" was Hermann Günther Grassmann in his book "Die lineare Ausdehnungslehre" (1844). But of course, the notion "vector space" was implicit in the work of several earlier authors. In Newton's "Principia" the velocities and forces are vectors. The addition of complex numbers was defined by Wessel and Argand as an addition of "directed line segments" in the plane. The Galois fields $GF(p^n)$ constructed by Galois are n-dimensional vector spaces over the prime field $GF(p)$. The algebra of quaternions is a four-dimensional vector space over $\mathbb{R}$, and Hamilton knew it, for he wrote in his letter to Graves: "And here dawned to me the notion that we must admit, in a sense, a *fourth dimension* of space...".

Grassmann's Ausdehnungslehre is very difficult to understand. His explanations are mixed with philosophical theories. His starting point is a "general doctrine of forms" which "ought to preceed all special branches of mathematics". His sums and products of vectors in an n-dimensional vector space are defined by purely geometrical considerations, without using basis elements $e_1, ..., e_n$. From our modern axiomatic point of view we can understand what he means, but his contemporaries did not understand it.

In 1862, Grassmann published another book entitled "Die Ausdehnungslehre vollständig und in strenger Form bearbeitet", but this book too had very little influence, as Grassmann notes himself in the "Vorrede" to the second edition (1878) of his first "Ausdehnungslehre".

However, as Grassmann states in the same Vorrede, the situation changed completely in 1867, when Hermann Hankel published Part 1 of his "Vorlesungen über die complexen Zahlen und ihre Funktionen" under the subtitle "Theorie der complexen Zahlensysteme". Chapter VII of this book is entitled "Theorie und geometrische Darstellung der alternierenden Zahlen". Hankel's "alternierende Zahlen" are just Grassmann's "Ausdehnungsgrößen", namely n-dimensional vectors and their alternating tensor products.

In the present section I shall follow Hankel's completely clear explanations, which are based, as Hankel himself says, on Grassmann's "Ausdehnungslehre von 1862", which I have not seen.

Hankel considers an algebra generated by elements $i_1, ..., i_n$ subject to the multiplication rules

$$i_k i_k = 0$$

$$i_k i_m = -i_m i_k.$$

He notes that if α and β are vectors:

$$\alpha = a_1 i_1 + \ldots + a_n i_n$$
$$\beta = b_1 i_1 + \ldots + b_n i_n$$

then the product

$$\alpha\beta = (a_1 b_2 - a_2 b_1) i_1 i_2 + (a_1 b_3 - a_3 b_1) i_1 i_3 + \ldots + (a_{n-1} b_n - a_n b_{n-1}) i_{n-1} i_n$$

has the property

$$\alpha\beta = -\beta\alpha.$$

More generally, a product of any number of vectors changes its sign if two successive factors are interchanged.

Following Grassmann, Hankel explains a geometrical interpretation of the alternating products of vectors. If two vectors lie in the same line, their product is zero. If not, they span a parallelogram lying in a certain plane and having a definite area. Two products ab and cd are equal if they lie in parallel planes and if they span parallelograms having the same area and the same sense of rotation from a to b as from c to d. Just so, a product of three vectors can be constructed as an oriented parallelepipedon, and so on.

Grassmann's products of vectors and alternating tensors are called *exterior products*. In modern presentations of the theory, for instance in Claude Chevalley's "Algebraic Theory of Spinors" (1954), Grassmann's algebra is enlarged by including a unit element. The algebra now has 2^n basis elements

$$1$$
$$i_a$$
$$i_{ab} = i_a i_b \qquad (a < b)$$
$$i_{abc} = i_a i_b i_c \qquad (a < b < c)$$
$$\vdots$$
$$i_{12\ldots n} = i_1 i_2 \ldots i_n.$$

In a historical appendix on p. 140 of his book, Hankel informs us that Grassmann's external multiplication has been rediscovered independently by Saint-Venant (1845), by O'Brien (1847), and by Cauchy (1853).

Clifford Algebras and Rotations in n Dimensions

In 1878, William Kingdon Clifford published a paper "Applications of Grassmann's Extensive Algebra", American Journal of Math. 1, p. 350–358, in which he defined an algebra generated by

$$1, i_1, \ldots, i_n$$

subject to the conditions

$$i_a^2 = -1, \quad i_a i_b = -i_b i_a.$$

As in the case of Grassmann's algebra, one can take as basis elements the 2^n elements

$$1$$

$$i_a$$

$$i_{ab} = i_a i_b \quad (a < b)$$

and so on. This algebra is called the *first Clifford algebra*. For $n=1$ one obtains the complex numbers, for $n=2$ the quaternions.

An important subalgebra is generated by the products of an even number of i's. Today this subalgebra is called the *second Clifford algebra*. For $n=3$ its basis elements

$$1, i_1 i_2, i_2 i_3, i_3 i_1$$

obey the multiplication rules of the quaternions.

In an extremely interesting, but nearly forgotten treatise "Untersuchungen über die Summen von Quadraten" (Bonn 1884), *R. Lipschitz* applied the second Clifford algebra to represent the rotations in n dimensions. He first showed that all rotations A, for which $\det(I+A)$ is not zero, can be obtained as

$$A = (I+C)^{-1}(I-C)$$

where C is an antisymmetric matrix. Next, putting

$$X = 1 x_1 + i_{12} x_2 + \ldots + i_{1n} x_n$$
$$Y = 1 y_1 + i_{12} y_2 + \ldots + i_{1n} y_n$$
$$\Lambda = 1 \lambda_0 + \Sigma i_{ab} \lambda_{ab} + \Sigma i_{abcd} \lambda_{abcd} + \ldots$$

(with $a < b < c < \ldots$), he proved that every rotation $X \to Y$ can be obtained from the formula

(8) $$\Lambda X = Y \Lambda_1,$$

in which Λ_1 is derived from Λ by replacing i_1 by $-i_1$.

This very remarkable formula can be written in a simpler form. Putting

$$X i_1 = x \quad \text{and} \quad Y i_1 = y,$$

and multiplying both sides of (8) by i_1, one gets

$$x = i_1 x_1 + \ldots + i_n x_n$$
$$y = i_1 y_1 + \ldots + i_n y_n$$

and

$$\Lambda x = y i_1^{-1} \Lambda_1 i.$$

But $i_1^{-1} \Lambda_1 i_1$ is just Λ, so we obtain the simple formula

(9) $\Lambda x \Lambda^{-1} = y.$

The formula (9) shows that there exists a representation of orthogonal transformations A by Clifford numbers Λ. The representation is not unique, for Λ can be multiplied by an arbitrary numerical factor, but Λ can be normed in such a way that only a factor ± 1 remains arbitrary. To the product of two Clifford numbers Λ and Λ' corresponds the product of the corresponding rotations, so we have here a two-valued representation of the group of rotations. The ground field may be $\mathbb{R}$ or $\mathbb{C}$.

The same two-valued representation of rotations in n dimensions was rediscovered for $n = 4$ by P.A.M. Dirac in 1928, and for all n by R. Brauer and H. Weyl in 1935. I shall now describe their approach.

Dirac's Theory of the Spinning Electron

P.A.M. Dirac's famous paper "The Quantum Theory of the Electron" was presented to the Royal Society in January 1928 and published in the Proceedings A 117, p. 610–629.

Dirac's starting point was a second-order relativistic wave equation for a free electron, which had been proposed by O. Klein and W. Gordon and others, and which can be written as:

(10) $(-p_0^2 + p_1^2 + p_2^2 + p_3^2 + m^2 c^2) \, \Psi = 0$

with

$$p_k = (h/2\pi i) \, \partial/\partial x_k$$

where x_1, x_2, x_3 are the space coordinates of the electron and $x_0 = ct$. Putting $p_0 = -i p_4$, Dirac rewrites (10) as

(11) $(p_1^2 + p_2^2 + p_3^2 + p_4^2 + m^2 c^2) \, \Psi = 0.$

For theoretical reasons, Dirac wants to replace (11) by a first-order wave equation of the form

(12) $\left(i \sum_1^4 \gamma_\mu p_\mu + mc \right) \Psi = 0.$

The equation (12) implies (11), provided the matrices γ_μ satisfy the conditions

(13)
$$\gamma_\mu^2 = 1$$
$$\gamma_\mu \gamma_\nu + \gamma_\nu \gamma_\mu = 0 \quad \text{for } \mu \neq \nu.$$

If these conditions are satisfied, the sum Σp_μ^2 becomes a complete square:

(14) $$p_1^2 + p_2^2 + p_3^2 + p_4^2 = (\Sigma \gamma_\mu p_\mu)^2.$$

Starting with Pauli's spin matrices

$$s_x = \begin{pmatrix} 0 & 1 \\ 1 & 0 \end{pmatrix}, \quad s_y = \begin{pmatrix} 0 & -i \\ i & 0 \end{pmatrix}, \quad s_z = \begin{pmatrix} 1 & 0 \\ 0 & -1 \end{pmatrix}$$

Dirac succeeds in constructing a set of 4×4-matrices γ_μ satisfying the conditions (13). Next he proves the invariance of the wave equation (12) by showing that every solution of (13) in 4×4-matrices may be obtained from the original solution by a transformation

(15) $$\gamma_\mu' = \tau \gamma_\mu \tau^{-1}.$$

If the p_μ are subjected to a Lorentz transformation

$$p_\mu' = \Sigma a_{\mu\nu} p_\mu$$

the left hand side of (14) remains invariant. The right hand side becomes

$$(\Sigma \gamma_\mu' p_\mu')^2$$

where the γ_μ' again satisfy the conditions (13), so that they may be written in the form (15). Thus, to every Lorentz transformation corresponds a 4×4-matrix τ, and one obtains a representation of the Lorentz group by 4×4-matrices τ.

In (15), the matrix τ can be replaced by $a\tau$, where a is an arbitrary constant. However, the matrices τ can be normed in such a way that only a factor ± 1 remains arbitrary, so we have here a two-valued representation of the Lorentz group.

If Dirac's matrices γ_μ are multiplied by i, one obtains matrices i_μ ($\mu = 1, 2, 3, 4$) satisfying the multiplication rules of the Clifford numbers. It follows that for $n = 4$ the Clifford algebra is isomorphic to a full matrix algebra over the field of complex numbers. We shall see presently that this is the case for all even values of n.

For the historical development of the quantum theory of the spinning electron I may refer to my article "Exclusion Principle and Spin" in the volume "Theoretical Physics in the Twentieth Century", edited by M. Fierz and V.F. Weisskopf, Interscience Publishers 1960.

Spinors in *n* Dimensions

In 1935, Richard Brauer and Hermann Weyl published a beautiful paper entitled "Spinors in *n* Dimensions", American Journal of Math. 57, p. 425–449, in which they used Clifford algebras in order to obtain two-valued matrix representations of the group of rotations in *n* dimensions.

The existence of such representations had been recognized as early as 1913 by Elie Cartan. In his Thèse de doctorat (1894) Cartan had developed a complete classification of all simple Lie algebras, and in his 1913 paper "Les groups projectifs qui ne laissent invariant aucune multiplicité plane" (Bulletin de la société math. de France 41, p. 53–96) the same Cartan determined all irreducible matrix representations of these Lie algebras. Among these he found a class of representations which, when integrated, led to two-valued representations of orthogonal groups.

The cases $n=3$ and $n=6$ were known already to Felix Klein and Sophus Lie. The real orthogonal group in 3 dimensions is locally isomorphic to the special unitary group $SU(2, \mathbb{C})$, and the real orthogonal group in 6 dimensions is locally isomorphic to the special unitary group $SU(4, \mathbb{C})$. If the ground field is extended to the complex number field $\mathbb{C}$, one obtains local isomorphisms to the special linear groups $SL(2, \mathbb{C})$ and $SL(4, \mathbb{C})$. For a full explanation of these and the related isomorphisms for $n=4$ and $n=5$ see my booklet "Gruppen von linearen Transformationen" (Springer-Verlag 1935) p. 18–28.

In his 1913 paper, Cartan restricted himself to describing the matrix representation of the *infinitesimal* transformations of the classical groups, including the rotation groups. On the other hand, Brauer and Weyl succeeded in constructing the *global* two-valued representations by means of Clifford algebras.

Right at the beginning of their paper, Brauer and Weyl state:

Our procedure is exactly the same as followed by DIRAC in his classical paper on the spinning electron. We introduce n quantities p_i which turn the fundamental quadratic form into the square of a linear form

$$x_1^2 + \ldots + x_n^2 = (p_1 x_1 + \ldots + p_n x_n)^2.$$

For this purpose we must have

$$p_i^2 = 1, \qquad p_k p_i = -p_i p_k \quad (k \neq i).$$

Clearly the quantities $p_1, \ldots, p_n$, multiplied by i, are just the Clifford numbers $i_1, \ldots, i_n$. Because of their relation with Dirac's theory of the spinning electron, Brauer and Weyl call their quantities "spinors". The same expression is also used by Elie Cartan in his two volumes "Leçons sur la théorie des spineurs" (Paris, Hermann 1938), in which he constructs two-valued representations of orthogonal groups by a geometrical method. This method, which is more complicated than that of Brauer and Weyl, will not be discussed here.

Brauer and Weyl first construct a matrix representation of the algebra of spinors. If n is even, $n=2\nu$, the algebra is isomorphic to a full matrix algebra of rank 2^ν. If n is odd, $n=2\nu+1$, the algebra of spinors is a direct sum of two matrix algebras, but it has a subalgebra, the second Clifford algebra generated by the even products of the p_i, which is a full matrix algebra of rank 2^ν.

If the matrices $P_1, \ldots, P_n$ representing the spinors $p_1, \ldots, p_n$ are transformed into $P_1', \ldots, P_n'$ by an orthogonal transformation of the P_k, one obtains an automorphism of the full matrix algebra. Now it is known (and proved in §11 of the paper of Brauer and Weyl) that every automorphism $X \to X^*$ of a full matrix algebra is an inner automorphism:

$$X^* = AXA^{-1}$$

in which the matrix A is uniquely determined but for a factor c. Thus one obtains a multi-valued representation of the orthogonal group. The matrices A can again be normed in such a way that only a factor ± 1 remains arbitrary.

For more details I may refer to the paper of Brauer and Weyl, which is also reprinted in the "Selecta Hermann Weyl" (Birkhäuser, Basel 1956), p. 431–454.

Chevalley's Generalization

In his booklet "The Algebraic Theory of Spinors" (Columbia University Press 1954) Claude Chevalley has generalized the theory of Clifford algebras. Instead of the sum $x_1^2 + \ldots + x_n^2$ from which Lipschitz as well as Brauer and Weyl started, Chevalley considers an arbitrary quadratic form $Q(u)$ defined on an n-dimensional vector space M. The ground field is completely arbitrary: it may even have characteristic two. From the vector space M and the form $Q(u)$ Chevalley constructs a generalized Clifford algebra.

In a paper "On Clifford Algebras", Proceedings Akademie Amsterdam A 69, p. 78–83, I have given a simplified account of Chevalley's construction. For the structure theory of Chevalley's generalized Clifford algebras and their application to the representation theory of orthogonal groups I may refer to the treatise of Chevalley.

If the quadratic form $Q(u)$ is identically zero, Chevalley's algebra becomes a Grassmann algebra.

Generalized Quaternions

Let F be a field of characteristic $\neq 2$. An algebra of *generalized quaternions* is generated by a basis $(1, i, j, k)$ and defined by the rules of multiplication

$$i^2 = \alpha, \qquad j^2 = \beta, \qquad k^2 = -\alpha\beta,$$

$$ij = k, \qquad jk = -i\beta, \qquad ki = -j\alpha,$$

$$ji = -k, \qquad kj = i\beta, \qquad ik = j\alpha.$$

The *norm* of a generalized quaternion

$$q = s + ix + jy + kz$$

is

(16)
$$N(q) = (s + ix + jy + kz)(s - ix - jy - kz)$$
$$= s^2 - \alpha x^2 - \beta y^2 + \alpha\beta z^2.$$

The algebra of generalized quaternions is semisimple in the sense of Maclagan Wedderburn (see Chapter 11). Hence it is either a division algebra or a full matrix algebra over the ground field F. More precisely: if the quadratic form (16) takes the value zero only for $s=x=y=z=0$, every non-zero q has an inverse q^{-1}, and the algebra is a division algebra. On the other hand, if the form takes the value zero, there are zero divisors and the algebra is a full matrix algebra of 2×2 matrices over F. So it is important to know under what conditions the quadratic form (16) takes the value zero.

Any quaternion can be written as

$$(17) \qquad q=(s+ix)+j(y-iz).$$

If z happens to be zero, we have

$$(18) \qquad q=s+ix+jy.$$

If z is not zero, we can multiply (17) on the right by $y+iz$, and thus obtain a quaternion of the form (18). The norm of this quaternion is

$$(19) \qquad N(q)=s^2-\alpha x^2-\beta y^2,$$

and we see: if the quaternary form (16) takes the value zero, so does the ternary form (19).

Now let F be the field of rational numbers. The conditions under which the diophantine equation

$$(20) \qquad s^2-\alpha x^2-\beta y^2=0$$

is solvable have been established in 1798 by Adrien-Marie Legendre (see his "Théorie des nombres", third edition, Vol. 1, §III and §IV). His condition can be formulated as follows:

In (20) one can assume α and β to be square-free integers. If δ is the largest common divisor of α and β, one can put $s=\delta \cdot t$ and divide the quation (20) by δ. Thus one obtains a reduced diophantine equation

$$(21) \qquad at^2+bx^2+cy^2=0,$$

in which a,b,c are integers such that abc is square-free.

Now the conditions for rational solvability of (21) are:

1) a,b,c are not all positive and not all negative.

2) If p is any odd prime factor of a (or of b or c), then $-bc$ (or $-ac$ or $-ab$) is a square residue modulo p.

It is clear that the conditions 1) and 2) are necessary. Now suppose they are satisfied. Legendre proves: If (21) were not solvable, one could construct another equation with smaller coefficients, also satisfying the conditions and not solvable. This would lead to an infinite descent, which is impossible. So (21) is solvable.

This method of "descente infinie" is due to Lagrange: "Sur la solution des problèmes indéterminés du second degrée", Mémoires de l'académie de Berlin 23 (1769)=Oeuvres de Lagrange II, p. 375-399. See also André Weil: Number Theory, An approach through history, pages 100-101 and 327-328.

In 1923, Helmut Hasse transformed the conditions 1) and 2) thus: The equation (21) is required to be solvable in real numbers not all zero, and also in p-adic numbers for all odd primes p occurring in the factorization of abc.

The general principle of which this theorem is a special case is called by Hasse "the local-global principle". It says: If a quadratic diophantine equation is locally solvable for all primes p, and also for the "locus" ∞, that is, if it is solvable in all p-adic fields and also in the field $\mathbb{R}$, it is solvable in rational numbers. This principle is valid not only for homogeneous quadratic equations in three unknowns, but quite generally for all homogeneous quadratic diophantine equations. See H. Hasse: Über die Darstellbarkeit von Zahlen durch quadratische Formen im Körper der rationalen Zahlen, Crelle's Journal für Math. 152, p. 129–148 (1923).

In a later paper (same Journal 153, p. 113–130, 1924) Hasse has generalized his local-global principle to arbitrary algebraic number fields. Here the p-adic number fields must be replaced by P-adic number fields defined by the prime ideals P of the number field F.

On the arithmetics of generalized quaternions see

B. Venkov: Zur Arithmetik der Quaternionen, Bulletin Acad. Sci. URSS (6) 16, p. 205–246 (1923),

C.G. Latimer: Arithmetics of Generalized Quaternion Algebras, Amer. Journal of Math. (2) 27, p. 92–102 (1926),

C.E. Wahlin: A Quadratic Algebra and its Application to a Problem in Diophantine Analysis, Bulletin Amer. Math. Soc. 33, p. 221–231 (1927),

I.W. Griffith: Generalized Quaternion Algebras and the Theory of Numbers, Amer. Journal of Math. 50, p. 303–314 (1928),

H. Brandt: Idealtheorie in Quaternionenalgebren, Math. Annalen 99, p. 1–29 (1928),

M. Eichler: Zur Zahlentheorie der Quaternionen-Algebren, Journal für Math. 195, p. 127–151 (1956),

G. Aeberli: Der Zusammenhang zwischen quaternären und quadratischen Formen und Idealen in Quaternionenringen, Commentarii Math. Helv. 33, p. 212–239 (1959).

Crossed Products

In 1929, in her lectures at Göttingen, Emmy Noether developed a theory of crossed products ("verschränkte Producte"). This theory was explained by Helmut Hasse in Part II of his 1932 paper "Theory of Cyclic Algebras over an Algebraic Number Field", Transactions American Math. Soc. 34, p. 180–200. See also Max Deuring: Algebren (Ergebnisse der Math. IV, 1, Springer-Verlag 1935), p. 52–67.

Let Z be a separable normal extension of a field F. Let n be the degree of Z over F and G its Galois group. A *crossed product* of G and Z is defined as an algebra A of the following type: A has a Z-basis consisting of n elements u_S corresponding to the elements S of G, for which the relations

(22) $$z u_S = u_S z^S$$

(23) $$u_S u_T = u_{S,T} a_{S,T}$$

hold, the $a_{S,T}$ being non-zero elements of Z. The set of coefficients $a_{S,T}$ is called the *factor set* of A. The associativity of A is ensured by the conditions

$$(24) \qquad\qquad a_{S,TR}\, a_{T,R} = a_{ST,R}\, a_{S,T}^{R}.$$

The algebra A is simple and normal over F, which means that its centre is F. The factor set may be replaced by an equivalent set

$$(25) \qquad\qquad b_{S,T} = a_{S,T}\, c_S^T\, c_T / c_{ST}.$$

The algebra obtained from A by extending the ground field F to a field L will be denoted by A_L. If A_L is a full matrix algebra, L is called a *splitting field* of A. Hasse proves on p. 184 of his paper that Z is always a splitting field.

Cyclic Algebras

An important class of crossed products is formed by the *cyclic algebras*. They were constructed by Leonard Eugen Dickson in his 1914 paper "Linear Algebras and Abelian Equations", Transactions American Math. Soc. 15, p. 31–46.

Let $Z = F(z)$ be a cyclic extension of degree n of the ground field F, and let

$$1, S, S^2, \ldots, S^{n-1} \qquad (S^n = 1)$$

be the automorphisms forming the Galois group of Z. As in the preceding section, the element resulting from z by applying the automorphism S will be denoted by z^S. The cyclic algebra A is now generated by the products

$$(26) \qquad\qquad u^i z^k \qquad (i, k = 0, 1, \ldots, n-1),$$

subject to the multiplication rules

$$(27) \qquad\qquad zu = uz^S$$

$$(28) \qquad\qquad u^n = \alpha,$$

where $\alpha \neq 0$ is a given element of F.

All products $z^k u^i$ can be reduced to products (26) by a repeated application of the rule (27). The result is

$$(29) \qquad\qquad z^k u^j = u^j (z^k)^{S^j},$$

and now, by applying (28) and (29), every product of two basis elements (26) can be expressed as a sum of terms (26) with coefficients from F.

If $n = 2$ and if the characteristic of the field F is not 2, we have a generalized quaternion algebra

$$u^2 = \alpha, \quad z^2 = \beta, \quad zu = -uz.$$

A cyclic algebra is always simple and normal over F. The algebra A is denoted by

$$A = (\alpha, Z, S).$$

According to Wedderburn, two cyclic algebras (α, Z, S) and (β, Z, S) over the same field F are isomorphic if and only if α/β is the norm of an element of Z. See J.H. Maclagan Wedderburn: On Division Algebras, Transactions American Math. Soc. 22, p. 129–135 (1921).

L.E. Dickson has proved: A is a division algebra if α^n is the least power of α which is the norm of an element of Z. On the other hand, if α itself is a norm of an element of Z, the algebra A is a full matrix algebra of $n \times n$-matrices.

Chapter 11
The Structure of Algebras

General Notions and Notations

The basis element of an algebra will be denoted by

$$e_1, \ldots, e_n.$$

Their multiplication is defined by

(1) $$e_j e_k = \sum e_i a_{ijk}$$

with coefficients a_{ijk} taken from a ground field F.

The elements of an algebra will be denoted like this:

(2) $$x = e_1 x_1 + \ldots + e_n x_n,$$

(3) $$u = e_1 u_1 + \ldots + e_n u_n,$$

the coefficients being either elements of F or indeterminates. If the coefficients $x_1, \ldots, x_n$ are indeterminates, the expression (2) will be called a *generic element* of the algebra.

With any element u of an associative algebra we may associate a linear transformation U:

$$x \to ux$$

of the algebra into itself. The mapping $u \to U$ is called the *regular representation* of the algebra. It has the usual properties of a representation of a ring:

$$u + v \to U + V$$
$$uv \to UV.$$

If E_j is the matrix representing e_j in the regular representation, we have

(4) $$U = E_1 u_1 + \ldots + E_n u_n.$$

The matrix elements of E_j are, for every fixed j, the elements a_{ijk}.

If the algebra has a unit element, the regular representation is an isomorphism, and the matrix X corresponding to a generic element x is non-singular.

As we have seen in the preceding chapter, every non-singular matrix in a matrix algebra has an inverse element within the algebra. It follows that the non-singular matrices U form an n-parametric Lie group. This important link between associative algebras and Lie groups was noted by Henri Poincaré in his paper "Sur les nombres complexes", Comptes Rendus Acad. Paris 99, p. 740–742 (1884). The infinitesimal transformations of this Lie group are just the matrices (4), and the Lie product $[U, V]$ is

(5) $$[U, V] = UV - VU.$$

A second Lie group is formed by the non-singular transformations

$$x \to xv.$$

Every transformation of the first Lie group commutes with every transformation of the second Lie group.

Benjamin Peirce

Benjamin Peirce's pioneer memoir "Linear Associative Algebras" was read before the National Academy of Science in Washington in 1870, and next lithographed in 100 copies for private circulation. It was published posthumously in 1881 in American Journal of Math. 4, p. 97–215, with addenda due to the author and his son C.S. Peirce.

The author's friend George Bancroft received a copy. In an accompanying letter, preserved in the manuscript division of the New York Public Library, Benjamin Peirce explained the main purpose of the memoir. The following extract from this letter is drawn from the article PEIRCE, BENJAMIN in Dictionary of Scientific Biography.

This work undertakes the investigation of all possible single, double, triple, quadruple, and quintuple Algebras which are subject to certain simple and almost indispensable conditions. The conditions are those well-known to algebrists by the terms of *distributive* and *associative*, which are defined on p. 21. It also contains the investigation of all sextuple algebras of a certain class, i.e. of those which contain what is called in this treatise an *idempotent* element.

Peirce was the first to introduce the notions *nilpotent* and *idempotent*. An element A is called nilpotent if some power A^n is zero, and it is called idempotent if $A^2 = A$.

If i is an idempotent element, every element A of the algebra can be written as a sum

(6) $$A = iA + (A - iA) = B + C,$$

the first term $iA = B$ having the property $iB = B$, while the term C has the property $iC = 0$.

The decomposition (6) is called in the modern literature a *right Peirce decomposition*. The elements $iA = B$ form a right ideal R having the property

$iR = R$, and the elements $A - iA = C$ form another right ideal R' having the property $iR' = 0$. The whole algebra is a direct sum of these two right ideals.

Just so, one can define a *left Peirce decomposition*. Combining the right and left decomposition, one obtains a *two-sided Peirce decomposition*

$$(7) \qquad A = iAi + (iA - iAi) + (Ai - iAi) + (A - iA - Ai + iAi).$$

These decompositions play an important part in the theory of Maclagan Wedderburn and in all later presentations of the general theory of algebras.

Peirce proved several theorems concerning nilpotent and idempotent elements. Next he investigated the possible types of algebras having at most 6 basis elements.

Eduard Study

Peirce's list of algebras of dimensions up to 6 was imcomplete, even for dimensions 3 and 4. A complete list of algebras with unity of dimensions up to 4 over the fields $\mathbb{R}$ and $\mathbb{C}$ was presented by Eduard Study: "Über Systeme von complexen Zahlen", Nachrichten Ges. der Wiss. Göttingen 1889, p. 237–268.

As we have seen in Chapter 10, Study used biquaternions to obtain a parametric representation of Euclidean motions.

The state of the theory of algebras in 1898 was summarized by Study in his article "Theorie der gemeinen und höheren complexen Größen" in the Encyklopädie der mathematischen Wissenschaften IA 4, Vol. 1, Part 1, p. 147–183. This article was brought up to date by Élie Cartan in 1908 in his article "Nombres complexes. Exposé d'après l'article allemand de E. Study" in the Encyclopédie des sciences mathématiques I, 5.

Gauss, Weierstrass and Dedekind

In a review in the "Göttinger gelehrte Anzeigen" for 1831 Carl Friedrich Gauss, speaking of complex numbers and their use in number theory, raised the question as to why "the relations between things which represent a multiplicity of more than two dimensions cannot yield other types of quantities permissible in general arithmetic" (see C.F. Gauss, Werke 2, p. 169–178).

In 1883, in a letter to Hermann Amandus Schwarz, Karl Theodor Weierstrass considered this question, giving it a concrete algebraic formulation. The letter was published in 1884 in the Nachrichten der Ges. der Wiss. Göttingen (p. 395–414) under the title "Zur Theorie der aus n Haupteinheiten gebildeten complexen Größen". In this paper, Weierstrass proved: A commutative algebra over $\mathbb{R}$ of dimension $n > 2$ always has divisors of zero.

In the next year 1885, Richard Dedekind published a paper under the same title in the same "Göttinger Nachrichten" (1885, p. 141–159), in which the problem was considered from a more general point of view. Dedekind starts with a commutative algebra over the complex number field defined by multiplication rules (1). He puts

(8)
$$p_{ij} = \sum_{r,s} a_{rsr} a_{sij}$$

and he assumes

(9)
$$\det(p_{ij}) \neq 0.$$

Under these conditions Dedekind shows that new basis elements e'_i can be introduced such that

(10)
$$e'_j e'_k = \begin{cases} 0 & \text{for } j \neq k \\ e'_k & \text{for } j = k. \end{cases}$$

This means: the algebra is a direct sum of copies of the complex field $\mathbb{C}$.

Georg Scheffers

As we have seen, Henri Poincaré pointed out in 1884 that every algebra over $\mathbb{R}$ or $\mathbb{C}$ having a unity element defines a pair of commuting Lie groups. This note made a strong impression on Sophus Lie, who in 1886 obtained a professorship at the University of Leipzig. In his seminary Lie repeatedly invited his listeners to apply the methods of Lie group theory to the investigation of the structure of algebras.

The suggestion was taken up by Lie's first pupil Georg Scheffers. In his paper "Zurückführung complexer Zahlensysteme auf typische Formen", Math. Annalen 39, p. 292–390 (1891), Scheffers considered associative algebras over the complex number field $\mathbb{C}$ having a unity element. His starting point was the connection between algebras and Lie groups just mentioned.

In the theory of Lie groups, an important distinction is made between *non-solvable* and *solvable* (or *integrable*) Lie groups. The latter have a composition series in which all factor groups are one-dimensional. Corresponding to this distinction, Scheffers introduces a distinction between *quaternion systems* and *non-quaternion systems*. By definition, an algebra is a non-quaternion system if it has a composition series of two-sided ideals of dimensions $1, 2, 3, \ldots, n$:

$$(e_1)$$
$$(e_1, e_2)$$
$$(e_1, e_2, e_3), \text{ etc.}$$

Scheffers now proves:

An algebra over $\mathbb{C}$ is a quaternion system if and only if it contains a subalgebra isomorphic to the algebra of quaternions and having the same unity element as the whole algebra.

An algebra A is called *reducible* if it is a direct sum of two proper subalgebras

$$A = B + C$$

such that $BC = 0$.

Scheffers proved several theorems concerning the structure of non-quaternion systems. These theorems enabled him to construct all irreducible non-quaternion systems having up to 5 basis elements.

By a different method, H. Rohr determined all algebras of dimensions up to 5 in his Ph D-theses (Marburg 1890) entitled "Über die aus 5 Haupteinheiten gebildeten complexen Zahlensysteme". His results agree with those of Scheffers.

In the continuation of his paper, Scheffers also determined all irreducible quaternion systems up to 8 basis elements.

Theodor Molien

Theodor Molien (1861–1941) was born in Riga and educated at the university of Dorpat (now Tartu) in Estonia. He spent a brief period (1884–85) at Leipzig, where Felix Klein was then a professor. In 1885 he became "Dozent" at the university of Dorpat, and in August 1891 he finished his thesis "Über Systeme höherer komplexer Zahlen", which was published in Math. Annalen 41, p. 83–156 (1893).

In 1897 Molien published two papers on the representations of finite groups, which will be discussed in a later chapter. In 1901 he became professor of mathematics at the university of Tomsk in Siberia.

I shall now summarize Molien's extremely important thesis of 1891. Molien considers associative algebras over the field $\mathbb{C}$ having a unity element. The product xu of two elements x and u is denoted by

$$(11) \qquad\qquad x'_i = \Sigma a^i{}_{jk} x_j u_k.$$

If this set of equations can be splitted into two partial systems of r and $n - r$ equations such that the first partial system contains only the variables $x_1, \dots, x_r$ and $u_1, \dots, u_r$:

$$(12) \qquad\qquad x'_i = \Sigma a^i{}_{jk} x_j u_k \qquad (i, j, k = 1, \dots, r)$$

then the algebra defined by (12) is called an *accompanying system* of the original algebra. In modern terminology the mapping

$$(x_1, \dots, x_n) \mapsto (x_1, \dots, x_r, 0, \dots, 0),$$

is a homomorphism of the algebra onto the accompanying system, and the equations

$$x_1 = 0, \dots, x_r = 0$$

define a two-sided ideal, the kernel of the homomorphism.

An algebra having no accompanying system with $0<r<n$ is a simple algebra in the modern sense. Molien calls it *primitive* ("ursprünglich").

In §3, Molien considers linear and bilinear forms on the algebra. If, in a linear form

$$g(x')=g_1 x'_1 +\ldots+g_n x'_n$$

the element x' is replaced by a product xu, one obtains a bilinear form

(13) $$g(xu)=\Sigma g_i a^i{}_{jk} x_j u_k,$$

which is called a *derived form* of the algebra. If the g_i are such that

(14) $$g(uv)=g(vu),$$

then the form (13) is said to have the *polar property*; I shall call it a *polar form*.

Molien notes that every algebra having a unity element has at least one non-zero polar form, namely the trace of xu in the regular representation:

(15) $$Tr(xu)= \sum_{i,j,k,s} a^i{}_{si} a^s{}_{jk} x_j u_k.$$

Note that the coefficients of this form are just the p_{jk} introduced by Dedekind in his investigation of commutative algebras:

(16) $$p_{jk}=\sum_{i,s} a^i{}_{si} a^s{}_{jk}.$$

Molien proves: If a derived form having the polar property has rank r, then an accompanying system of dimension r exists. It follows that a simple algebra has only one polar form, namely the form (15), and that this form has rank n. Conversely: If the form (15) has rank n and if it is the only derived polar form, the algebra is simple.

At the end of Chapter II, Molien proves a theorem of fundamental importance, namely:

Theorem 30. The basis of any simple algebra can be chosen in such a way that the product equations assume the form

(17) $$x'_{ik}=\sum_{j} x_{ij} u_{jk}.$$

This means, in modern terminology:
Every simple algebra over $\mathbb{C}$ is a complete matrix algebra.

As we shall see later, this theorem is a special case of a fundamental theorem of Maclagan Wedderburn.

For a more detailed analysis of Molien's thesis I may refer to a very interesting paper by Th. Hawkins: Hypercomplex Numbers, Lie Groups, and the Creation of Group Representation Theory, Archive for History of Exact Sciences 8, p. 243–287 (1972). Hawkins shows that the work of Molien is closely related to that of Killing on Lie algebras.

Élie Cartan

Several results obtained by Molien in 1891 and published by him in 1893 were rediscovered by Élie Cartan and published in 1898 in a paper entitled "Les groupes bilinéaires et les systèmes de nombres complexes", Annales de la Faculté des Sciences de Toulouse 12 B, p. 1–99 (Oeuvres complètes, Partie 2, Vol. 1, p. 7–105).

It seems that Cartan was not aware of Molien's work, for the earliest mention of Molien in Cartan's publications is in his article "Nombres complexes" in the Encyclopédie des sciences mathématiques, which appeared in 1908. In this article, Cartan fully acknowledges Molien's accomplishments.

Cartan's methods of proof are quite different from those of Molien. In his Thèse of 1894, Cartan had investigated the structure of Lie algebras, in particular of simple and semi-simple Lie algebras. In his 1898 paper just mentioned, he applied the same methods to the study of algebras, in particular of simple and semi-simple algebras. The ground field is always $\mathbb{C}$, and the algebra Σ is always supposed to have a unity element.

Cartan starts with the regular representation, in which every element of Σ

$$x = e_1 x_1 + \ldots + e_r x_r$$

is represented by the linear transformation

$$y \to x y.$$

In the regular representation, a generic element x of the algebra Σ is represented by a matrix

(18) $$X = E_1 x_1 + \ldots + E_r x_r.$$

The *characteristic equation* of the matrix X is

(19) $$\det(X - \omega I) = 0.$$

Let $\omega_1, \omega_2, \ldots, \omega_h$ be the *different* roots of this equation. Cartan now replaces the generic element x by an element a such that $\omega_1, \ldots, \omega_h$ are all different.

Consider one of the roots, say ω_1. If the root has multiplicity $m_1 = m$, there exists a sequence of elements

$$\alpha_1, \alpha'_1, \alpha''_1, \ldots, \alpha_1^{(m-1)}$$

such that

$$a\alpha_1 = \omega_1 \alpha_1$$
$$a\alpha'_1 = \omega_1 \alpha'_1 + \lambda_{11} \alpha_1$$
$$a\alpha''_1 = \omega_1 \alpha''_1 + \lambda_{21} \alpha_1 + \lambda_{22} \alpha'_1 \text{ etc.}$$

Thus, every root defines a linear subspace generated by the elements $\alpha_1, \alpha_1', \alpha_1'', \ldots$ such that, if y belongs to the subspace, so does yz for every z in the algebra. In other words, the subspaces are right ideals, and the whole algebra is a direct sum of these right ideals.

In particular, the unity element ε can be written as a sum of elements ε_i belonging to the h subspaces:

$$\varepsilon = \varepsilon_1 + \varepsilon_2 + \ldots + \varepsilon_h$$

and we have

$$\varepsilon_i^2 = \varepsilon_i, \quad \varepsilon_i \varepsilon_j = 0 \quad \text{for } i \neq j.$$

An element η is said to have the character (α, β) if

(20) $$\varepsilon_\alpha \eta = \eta \varepsilon_\beta = \eta,$$

which means that η belongs to the right ideal generated by ε_α and to the left ideal generated by ε_β. These characters (α, β) had been introduced already by Scheffers. The conditions (20) imply

$$\varepsilon_i \eta = 0 \quad \text{for } i \neq \alpha$$
$$\eta \varepsilon_j = 0 \quad \text{for } j \neq \beta.$$

The product of an element of character (α, β) by an element of character (γ, δ) is zero if $\beta \neq \gamma$, and it has character (α, δ) if $\beta = \gamma$.

Next, Cartan investigates the properties of nilpotent elements. He calls them "nombres pseudo-nuls", and defines them by the property that their characteristic roots are all zero. This definition is equivalent to that of Peirce.

Cartan's theory culminates in two theorems. The first theorem says, in modern terminology:

Every simple algebra over $\mathbb{C}$ is a full matrix algebra.

This theorem had been obtained by Molien, but Dickson notes that Molien's proof is not altogether satisfactory. See L.E. Dickson: Linear Algebras (Cambridge Tracts 16, New York 1914), p. 56.

Cartan defines a *semi-simple* algebra as a direct sum of simple algebras. His second theorem says:

Every algebra over $\mathbb{C}$ is a direct sum of a simple or semi-simple subalgebra and a nilpotent invariant subalgebra (i.e. a nilpotent two-sided ideal).

Maclagan Wedderburn defined a semi-simple algebra as an algebra having no nilpotent invariant subalgebra. Cartan's second theorem implies that a semi-simple algebra in the sense of Wedderburn is also semi-simple in the sense of Cartan, and conversely.

Next, Cartan studies the structure of simple and semi-simple algebras over the real number field $\mathbb{R}$. Once more, his theorems can be obtained as special cases of those of Wedderburn.

Maclagan Wedderburn

The first to develop a general theory of algebras over an arbitrary field was J.H. Maclagan Wedderburn. His paper "On Hypercomplex Numbers" was published in 1908 in Proc. London Math. Soc. (2) 6, p. 77–118.

Wedderburn defines a "complex" as a linear subspace of the algebra under consideration. He defines the sum and the product of two complexes. A subcomplex B of an algebra A having the properties

$$AB \subseteq B \quad \text{and} \quad BA \subseteq B$$

is called an *invariant* subcomplex of A. We call it a two-sided ideal. Theorem 3 says that every such ideal defines a residue-class algebra A/B. Wedderburn calls it the *difference algebra* of A and B.

Obviously, every algebra A_1 has a composition series of invariant sub-algebras

$$A_1 \supset A_2 \supset \dots .$$

Wedderburn now proves that the residue algebras

$$A_r/A_{r+1}$$

are unique but for their order.

If A is a sum $B + C$ of two subalgebras, and

$$BC = 0 = CB,$$

A is said to be *reducible*.

Next, Wedderburn proves several theorems concerning nilpotent subal-gebras, culminating in

Theorem 13. If N is a maximal nilpotent invariant subalgebra of an algebra A, all nilpotent invariant subalgebras are contained in N.

The uniquely defined subalgebra N is what we today call the *radical* of A. If N is zero, A is called *semi-simple*. Wedderburn's *Main Theorem* says:

1. Any algebra is the sum of its radical N and a semi-simple algebra.
2. A semi-simple algebra can be uniquely expressed as a direct sum of simple algebras.
3. A simple algebra is a full matrix algebra over a division algebra.

In his proofs, Wedderburn makes an extensive use of the methods of Peirce. He also quotes Scheffers, Molien, and Cartan.

Emil Artin

In 1927, Wedderburn's main theorem was generalized by Emil Artin in a paper "Zur Theorie der hyperkomplexen Zahlen", Abhandlungen math. Seminar Hamburg 5, p. 251–260. Instead of algebras over a field, Artin considered rings satisfying a "descending chain condition", which says: Every descending chain of left (or right) ideals

$$I_1 \supset I_2 \supset \dots$$

is finite. Chain conditions of this kind has been introduced by Emmy Noether in her theory of ideals and in her lectures on group theory. Assuming a descending chain condition for left or right ideals, Artin was able to prove Wedderburn's main theorem for simple and semi-simple rings.

Emmy Noether and her School

In 1924, when I came to Göttingen as a student, I had the pleasure to attend a course of Emmy Noether on Hypercomplex Numbers. In 1926/27 she again lectured on the same subject. This time the title of her course was "Hyperkomplexe Zahlen und Darstellungstheorie". My lecture notes are lost, but the contents of my notes were incorporated in Emmy Noether's 1929 paper "Hyperkomplexe Größen und Darstellungstheorie", Math. Zeitschrift 30, p. 641–692. Moreover, the Chapters 2 and 3 of Max Deuring's book "Algebren" (Springer 1935) are mainly based on Emmy Noether's lectures.

In the school of Emmy Noether, the structure theory of Wedderburn and Artin was extended to more general rings. Gottfried Köthe defined: A right or left ideal N in a ring A is called a *nilideal*, if all its elements are nilpotent. The union of all two-sided nilideals is a two-sided nilideal; Köthe calls it the *radical R* of A. See Gottfried Köthe: "Die Struktur der Ringe, deren Restklassenring nach dem Radikal vollständig reduzibel ist", Math. Zeitschrift 32, p. 161–186 (1930).

In the school of Emmy Noether it was felt that the radical R, as defined by Wedderburn or by Köthe, is too small. If the descending chain condition is not satisfied, it was not possible to obtain satisfactory structure theorems for A/R. Therefore, Reinhold Baer and Jakob Levitzki have proposed other definitions of the radical. See:

R. Baer: Radical Ideals, American Journal of Math. 65, p. 537–568 (1943).

J. Levitzki: On the Radical of a General Ring, Bulletin American Math. Soc. 49, p. 462–466 (1943).

However, these definitions still did not lead to a satisfactory structure theory of rings without finiteness conditions. The first to solve the riddle and to present a beautiful general theory was Nathan Jacobson.

Nathan Jacobson

In two papers published in 1943, Jacobson developed a structure theory for rings without finiteness assumptions. The titles are:

Structure Theory of Simple Rings without Finiteness Assumptions, Trans. Amer. Math. Soc. 57, p. 228–245 (1945).

The Radical and Semi-Simplicity for Arbitrary Rings, American Journal of Math. 67, p. 300–320 (1945).

Let A be a ring, and M a right A-module such that $MA \neq 0$. The module is called *simple* or *irreducible* if it has no proper A-submodule other than the zero module. To every element a of A there corresponds a linear transformation of M into itself

$$x \mapsto xa,$$

and these transformations define a representation of A. The elements a represented by zero form a two-sided ideal: the kernel of the representation induced by the module M.

Now the *radical* R of A is defined as the intersection of all kernels of irreducible representations, that is, of representations induced by simple A-modules.

This is an external definition, because it uses the totality of all irreducible A-modules. However, Jacobson proves that it can be replaced by an internal definition.

A right ideal I in A is called *modular* if there is an element e in A such that

$$ea \equiv a\,(I) \quad \text{for all } a \text{ in } A.$$

Jacobson proves that every irreducible A-module M is isomorphic to a residue class module A/I, where I is a maximal modular right ideal. It follows that the radical R is the intersection of the maximal modular right ideals.

Jacobson also proves that the radical contains all one-sided nilideals. So the radical of Köthe is contained in the radical of Jacobson.

The ring A is called *semi-simple* if its Jacobson radical is zero. For an exposition of Jacobson's theory I may refer the reader to Jacobson's excellent book "Structure of Rings", American Math. Soc. Colloquium Publ. 37, 1956.

Normal Simple Algebras

As we have seen, every simple algebra A is a full matrix algebra over a division algebra D. The centre of D is a finite extension of the ground field F. If D is *normal*, that is, if its centre is F, the full matrix algebra A is also normal.

The direct product $A \times B$ of two algebras A and B with basis elements $u_1, \ldots, u_m$ and $v_1, \ldots, v_n$ is defined as an algebra with basis elements

$$u_j v_k \quad (j = 1, \ldots, m; \ k = 1, \ldots, n).$$

The v are assumed to commute with the u, so a product

$$(u_h v_i)(u_j v_k)$$

can be computed as

$$(u_h u_j)(v_i v_k).$$

The direct product of two simple algebras $A \times B$ is again simple, provided the product $Z \times Z'$ of their centres is simple. Hence, if A and B are both normal simple algebras, their product is normal and simple. If A is a full matrix ring over a division algebra D, and B a full matrix ring over D', the product $A \times B$ is a full matrix ring over $D \times D'$.

The *Brauer Group* is defined as follows. Two normal simple algebras over F are said to be in the same class, if they are isomorphic to full matrix rings over one and the same division algebra D. Thus, every class $\{D\}$ belongs to just one division algebra D. If A and B are full matrix rings over D and D', the product belongs to $\{D \times D'\}$.

To every algebra A we can define an *inverse isomorphic* algebra A^* as follows:

$$ab = c \quad \text{implies} \quad b^*a^* = c^*.$$

Now one can prove that $A \times A^*$ is a full matrix algebra over F, so the class $\{D\} \times \{D^*\}$ is the unit class $\{F\}$. It follows that the classes $\{D\}$ form a commutative group, the *Brauer Group* belonging to the field F. See Richard Brauer: Über Systeme hyperkomplexer Zahlen, Math. Zeitschrift 30, p. 79–107 (1929).

If the ground field F is extended to S, the composition constants of A remaining unchanged, one obtains an enlarged algebra

$$A_S = A \times S.$$

If A is normal and simple, so is A_S: its centre is S. If A_S is a complete matrix ring over S, the field S is called a *splitting field*. A splitting field of a division algebra D is a splitting field of the whole class $\{D\}$.

According to A.A. Albert and Emmy Noether, splitting fields of D which are finite extensions of F can be characterized as those fields that can be imbedded, by irreducible matrix representations, as maximal subfields into algebras A of the class $\{D\}$. In particular, all maximal subfields of D are splitting fields. See A.A. Albert: On direct products, Transactions Amer. Math. Soc. 33, p. 690–711 (1931), and E. Noether: Nichtkommutative Algebren, Math. Zeitschrift 37, p. 514–541 (1933).

Gottfried Köthe has proved (Journal für Math. 166, p. 182–184) the existence of *separable* splitting fields which are maximal subfields of D.

Since splitting fields exist, the dimension $(D:F)$ is always a square. This was first proved by J.H. Maclagan-Wedderburn in his fundamental paper "On hypercomplex numbers", Proc. London Math. Soc. (2) 6, p. 77–118 (1908).

For proofs and fuller references I may refer to M. Deuring: Algebren (Springer-Verlag 1934), p. 46–47.

The Structure of Division Algebras

Since all simple algebras are full matrix rings over division algebras D, the problem to find all simple algebras over a given field F is equivalent to the problem of finding all division algebras over F. This problem can be solved for several important types of fields F, namely

1. The field $\mathbb{R}$
2. Finite fields
3. P-adic fields
4. Algebraic number fields.

I shall now consider these four cases separately.

1. Division Algebras over $\mathbb{R}$

In this case, the only possible division algebras are

the real number field $\mathbb{R}$,

the complex number field $\mathbb{C}$,

the algebra of quaternions.

This was proved by Georg Frobenius in 1878 in a paper "Über lineare Substitutionen und bilineare Formen", Crelle's Journal für Math. 84, p. 1–63. See also Olive Hazlett: On the Theory of Associative Division Algebras, Trans. Amer. Math. Soc. 18, p. 167–176 (1917).

2. Finite Skew Fields

Division algebras over a prime field $GF(p)$ are finite skew fields. They are all commutative, that is, they are the well-known Galois fields $GF(p^n)$. This was first proved by J.H. Maclagan Wedderburn: A Theorem on Finite Algebras, Trans. Amer. Math. Soc. 6, p. 349–352 (1909). Other proofs were given by L.E. Dickson (1927), E. Artin (1928), R. Brauer (1929), and E. Witt (1931). For references see M. Deuring: Algebren, p. 49 and 65.

3. Normal Simple Algebras over a P-adic Field

A P-adic field F_P is defined by a prime ideal P in an algebraic number field F, as follows. If π is an integer divisible by P, but not by P^2, the elements of F_P are formal power series

$$u = c_{-n}\pi^{-n} + \ldots + c_{-1}\pi^{-1} + c_0 + c_1\pi + \ldots,$$

the coefficients c_i being quotients of algebraic integers a/b, in which the denominator b is not divisible by P. These "P-adic numbers" were introduced by Kurt Hensel in 1904, and used in the theory of algebraic number fields by Hensel and Hasse. See

Kurt Hensel: Theorie der algebraischen Zahlen I, Leipzig 1908, and

Helmut Hasse: Zahlentheorie, third edition 1969; English translation: Number Theory, Springer-Verlag 1980.

The structure of division algebras over a P-adic field F_P was investigated by Helmut Hasse in a sequence of papers 1931–1933. For full references and proofs see M. Deuring: Algebren, p. 109-114. The main results of Hasse's theory may be summarized as follows:

Let F_p be a P-adic field, and let R be its ring of integers. The element π generates a prime ideal P in R. The residue class ring R/P is a $GF(q)$. For every fixed integer f, let w be a primitive $(q^f - 1)$th root of unity, and let

$$W_f = F_p(w)$$

be the extension of F_p generated by w. Then W_f is the only unramified extension of F_p of degree f, and every normal simple algebra A of dimension f^2 over F_p is a cyclic algebra

$$A = (\alpha, W_f, S)$$

where S is the automorphism of the field W_f defined by

$$w \mapsto w^q.$$

If α is divisible by π^r, but not by π^{r+1}, then the Brauer class of the algebra A is completely determined by the fraction r/f modulo 1. Thus, the Brauer Group is isomorphic to the additive group of rational integers r/f modulo 1.

4. Division Algebras over an Algebraic Number Field

Chapter VII of Deuring's "Algebren" is wholly devoted to this subject. The main contributors to this theory are A.A. Albert, Richard Brauer, Helmut Hasse, and Emmy Noether. I shall now summarize their results.

Let F be an algebraic number field, and let $Z = F(z)$ be a cyclic extension of F, of degree $(Z:F) = n$. If S is an automorphism generating the Galois Group of Z, we have defined a cyclic algebra

$$A = (\alpha, Z, S)$$

as an algebra over F having basis elements

$$u^i z^k \qquad (i, k = 0, 1, \ldots, n-1)$$

and multiplication rules

$$z u = u z^S$$
$$u^n = \alpha \qquad (\alpha \text{ in } F).$$

In Chapter 10 we have seen that A is a full matrix ring over F if and only if α is a norm of an element of Z over F.

Now the question, under what conditions α is a norm in Z can be answered by class field theory. Let me first recall a few notions from valuation theory.

According to David van Dantzig, every valuation V of a field F defines a complete extension F_V of F, in which the convergence criterion of Cauchy holds. See D. van Dantzig: Zur topologischen Algebra I, Math. Annalen 107, p. 587–626 (1932).

The valuations of algebraic number fields have been completely determined by Alexander Ostrowski in his fundamental sequence of papers "Untersuchungen zur arithmetischen Theorie der Körper", Math. Zeitschrift 39, p. 296–404. The main results of this paper, as far as they are concerned with algebraic number fields, may be summarized as follows.

If F is an algebraic number field, there are two kinds of valuations, namely P-adic valuations defined by prime ideals P of F, and archimedean valuations defined by the possible embeddings of F into the real number field $\mathbb{R}$ or into the complex number field $\mathbb{C}$. In the latter case, the field F_V is just $\mathbb{R}$ or $\mathbb{C}$, and the valuation is defined by the classical absolute value:

$$\varphi(a) = |a| \quad \text{for every element } a \text{ of } \mathbb{R} \text{ or } \mathbb{C}.$$

Now let Z be a cyclic extension field of F. Every valuation V of F can be extended to a valuation of Z. In the P-adic case, since Z is cyclic, all prime ideal factors of P in Z are transformed into each other by the automorphisms of the Galois group, so we may choose one prime factor at random and call the corresponding valuation V: it does not matter which prime factor we choose. In the archimedean case all embeddings of F into $\mathbb{R}$ or $\mathbb{C}$ are obtained from one embedding by applying the Galois group, so in this case too we may choose an extension of the valuation V at random and call it V. In any case the complete field Z_V is an extension of the complete field F_V.

Now the condition for an element α of F to be a norm of an element of Z is, that for every valuation V of F the element α of F_V be a norm of an element of Z_V. This is Hasse's "local-global principle" for cyclic extensions Z of a number field F.

Let us see what this means for P-adic and for archimedean valuations. In the P-adic case α is a norm of an element of Z_V if and only if the norm-residue symbol

$$\left(\frac{\alpha, Z}{P}\right)$$

is equal to 1. For an explanation of the meaning of this symbol see Max Deuring: Algebren, p. 123–126.

In the archimedean case the fields F_V and Z_V are either $\mathbb{R}$ or $\mathbb{C}$. If both are $\mathbb{R}$ or if both are $\mathbb{C}$, there is no condition for α, for in these cases every element of F_V is trivially a norm of an element of Z_V. But if $F_V = \mathbb{R}$ and $Z_V = \mathbb{C}$, there is a condition. The norm of a non-zero element a of $\mathbb{C}$ with respect to $\mathbb{R}$ is $a\bar{a}$, so it is always positive. Hence the required condition is, that in all embeddings of F into $\mathbb{R}$ the map of α is always positive, or in other words, that all real conjugates of the algebraic number α be positive.

An immediate consequence of the local-global theorem is:

The cyclic algebra A is a complete matrix ring over F if for every valuation V of F the extended algebra

$$A_V = A \times F_V$$

is a complete matrix ring over F_V.

This "local-global theorem" for cyclic algebras is a generalization of the same theorem for generalized quaternion algebras (see Chapter 10). It is due to Helmut Hasse (Transactions Amer. Math. Soc. 34, p. 171–214, 1932), but the idea to connect the theory of cyclic algebras with the norm-residue theory is due to Emmy Noether. In this connection I may quote Hermann Weyl's Memorial Address "Emmy Noether" (Scripta mathematica 3, p. 201–220, 1935):

> Hasse acknowledges that he owed the suggestion for his beautiful papers on the connection between hypercomplex quantities and the theory of class fields to casual remarks by Emmy Noether. She could just utter a far-seeing remark like this, "Norm rest symbol is nothing else than cyclic algebra" in her prophetic lapidary manner, our of her mighty imagination that hit the mark most of the time and gained in strength in the course of years; and such a remark could then become a signpost to point the way for difficult future work.

In a classical paper entitled "Beweis eines Hauptsatzes in der Theorie der Algebren" (Journal für Math. 167, p. 399–404, 1932), Helmut Hasse, Richard Brauer, and Emmy Noether generalized the local-global theorem to all normal simple algebras A over an algebraic number field F. They proved:

A is a full matrix ring over F if and only if all algebras A_V are full matrix rings over F_V.

Two proofs of this beautiful theorem can be found in Deuring's "Algebren", p. 117 and 132. The second proof is due to Max Zorn: Note zur analytischen hyperkomplexen Zahlentheorie, Abhandlungen math. Seminar Hamburg 9, p. 197–201 (1933).

Basing themselves on this local-global theorem, Hasse, Brauer, and Noether were able to obtain their "Main Theorem":

Every normal division algebra over an algebraic number field F is a cyclic algebra over F.

Chapter 12
Group Characters

The present chapter will be divided into two parts:
Part A: Characters of Abelian Groups
Part B: Characters of Finite Groups

Part A
Characters of Abelian Groups

Genera and Characters of Quadratic Forms

The history of the theory of group characters begins with Gauss. In Sections 228–233 of his "Disquisitiones arithmeticae", Gauss discusses the question: What kind of integers n can or cannot be represented by a given binary quadratic form

$$(1) \qquad\qquad F = ax^2 + 2bxy + cy^2$$

with integer coefficients a, b, c?

Gauss restricts himself to *primitive forms*, that is, he assumes a, b, c to have no common divisor. He defines the *determinant* D of the form F as

$$(2) \qquad\qquad D = b^2 - ac$$

and he proves:

If p is an odd prime divisor of D, the numbers n not divisible by p that can be represented by F are either all residues, or all are non-residues modulo p.

If the form is primitive and if p divides D, then a and c cannot both be divisible by p. Suppose a is not. Depending on whether a is a residue or not, the form will represent only residues or non-residues.

If the odd prime p is replaced by 2, the situation is more complicated. We have to take into account not only the modulus 2, but also its powers 4 and 8. Gauss proved:

If $D \equiv 3 \pmod 4$, the odd numbers represented by F are either all $\equiv 1$ or all $\equiv 3 \pmod 4$.

If $D \equiv 2 \pmod 8$, the odd numbers represented by F are either all $\equiv 1$ or $\equiv 7$, or all $\equiv 3$ or $\equiv 5 \pmod 8$.

If $D \equiv 6 \pmod 8$, the odd numbers represented by F are either all $\equiv 1$ or $\equiv 3$, or all $\equiv 5$ or $\equiv 7 \pmod 8$.

If $D \equiv 4 \pmod 8$, the odd numbers represented by F are either all $\equiv 1$, or all $\equiv 3 \pmod 4$.

If $D \equiv 0 \pmod 8$, the odd numbers represented by F are either all $\equiv 1$, or all $\equiv 3$, or all $\equiv 5$, or all $\equiv 7 \pmod 8$.

Gauss now defines, for every form F, certain "characters", as follows:

If the form F represents only residues modulo p, Gauss assigns to F the character $R \cdot p$, and if it represents only non-residues, he writes $N \cdot p$. Similarly he writes $1, 4$ if F represents no other odd numbers than those $\equiv 1 \pmod 4$, and so on. For instance, the complete character of the form

$$F = 7x^2 + 23y^2$$

will be

$$3, 4; \quad N \cdot 23; \quad R \cdot 7,$$

which means that a and c and hence all odd numbers represented by the form are $\equiv 3 \pmod 4$, that a is a non-residue mod 23, and that c is a residue mod 7.

In 1839, Lejeune Dirichlet used the Legendre symbol $\left(\dfrac{n}{p}\right)$, which is $+1$ for residues and -1 for non-residues modulo p, in order to write the results of Gauss in a more convenient form. His theorems read:

Theorem 2.1. If p is an odd prime divisor of D, then $\left(\dfrac{n}{p}\right)$ remains constant for all n not divisible by p.

Theorem 2.2. If $D \equiv 3 \pmod 4$, $(-1)^{(n-1)/2}$ remains constant for all odd n.

Theorem 2.3. If $D \equiv 2 \pmod 8$, $(-1)^{(n^2-1)/8}$ remains constant for all odd n.

Theorem 2.4. If $D \equiv 6 \pmod 8$, $(-1)^{(n-1)/2 + (n^2-1)/8}$ remains constant for all odd n.

Theorem 2.5. If $D \equiv 4 \pmod 8$, $(-1)^{(n-1)/2}$ remains constant for all odd n.

Theorem 2.6. If $D \equiv 0 \pmod 8$, both $(-1)^{(n-1)/2}$ and $(-1)^{(n^2-1)/8}$ remain constant for all odd n.

Two forms F and G belong to the same *class*, if there is an integer linear transformation

$$x = ax' + by'$$
$$y = cx' + dy'$$

with $ad - bc = 1$ transforming F into G. It is clear that two forms belonging to the same class represent the same numbers. It follows that the characters defined by Gauss depend only on the class to which the form F belongs.

Forms with negative determinants are either positive or negative. Forms with positive determinants are indefinite: they take positive as well as negative values. Now consider two forms F and G which are either both positive or both negative or both indefinite. Gauss defines: F and G belong to the same *genus*, if they have the same determinant and the same characters.

Gauss next defines a *composition* of forms, or rather of form classes. According to Gauss, the classes of primitive forms having a given determinant form what we now call a *finite abelian group*.

If the form F represents a number n, and if G represents n', any form of the product class will represent nn'. Now if n and n' are both residues or both non-residues modulo p, the product nn' will be a residue, and if only one of the two is a residue, the product is a non-residue. Similarly, if n and n' are both $\equiv 1$ (mod 4), the product will also be $\equiv 1$ (mod 4), and so on. So, if the characters of two classes of forms are known, the characters of the product class (mod p or mod 4 or mod 8) are known.

Basing himself on these results of Gauss, Peter Gustav Lejeune Dirichlet has given the following definition:

Any one of the symbols

(3) $$\left(\frac{n}{p}\right), \quad (-1)^{(n-1)/2}, \quad (-1)^{(n^2-1)/8}$$

used in the Theorems 2.1 to 2.6, and also any finite product of some of these symbols, is called a character of a class g of forms G. These characters are functions $\chi(g)$ having the property

(4) $$\chi(gg') = \chi(g)\,\chi(g').$$

For any given determinant $D \neq 0$, we have only a finite number of primes dividing D, and hence only a finite number, say λ, of symbols (3) defining a genus. The number of possible values of the set of characters (3) is 2^λ, for every character can take the value $+1$ or -1. However, Gauss has proved that only half of the possible sets of characters are realized by form classes. Thus, the number of genera is only $2^{\lambda-1}$.

Dirichlet's version of Gauss' theory of genera was first published in 1839 in a paper entitled "Recherches sur diverses applications de l'analyse infinité-simale à la théorie des nombres", Crelle's Journal für Math. 19, p. 324–369. The contents of this paper was expounded by Richard Dedekind in the fourth supplement to Dirichlet's "Vorlesungen über Zahlentheorie" (third edition, 1879).

The notion "character" was generalized by Dedekind in 1879 in his famous "tenth supplement" to Dirichlet's Vorlesungen. Dedekind considers functions $\chi(A)$ of ideals A in an algebraic number field having the property

(5) $$\chi(AB) = \chi(A)\,\chi(B)$$

and depending only on the class of the ideal A. Now the classes of ideals in an algebraic number field form a finite abelian group. So the characters $\chi(A)$ are group characters of an abelian group. Dedekind realizes this from the very beginning, for in a letter to Frobenius he said:

After all this (i.e. after Dirichlet's investigations) it was not much to introduce the concept and name of characters for every Abelian group, as I did in the third edition of Dirichlet's Zahlen-theorie.

The ideas of Dedekind on characters of abelian groups were explained in greater detail by his friend Heinrich Weber in 1882 in a paper "Beweis des Satzes, daß jede eigentlich primitive quadratische Form unendlich viele Primzahlen darzustellen fähig ist" (Math. Annalen 20, p. 301–329). In this paper, Weber defines finite abelian groups by means of postulates, gives a new proof of the "fundamental theorem on abelian groups", and shows how all characters of such a group can be obtained from the characters of the fundamental generators $\theta_1, \theta_2, \ldots$ of orders $n_1, n_2, \ldots$. One just assigns to each of the generators θ_i and n_i-th root of unity ω_i and one puts

$$\chi(\theta_1^{h_1} \theta_2^{h_2} \ldots) = \omega_1^{h_1} \omega_2^{h_2} \ldots .$$

The product of two characters is again a character. The characters form an abelian group isomorphic to the original group.

In a nutshell, the theory of characters of finite abelian groups is already contained in Dedekind's "tenth supplement".

Duality in Abelian Groups

We have defined characters χ of a finite abelian group G by the property

(6) $$\chi(ab) = \chi(a)\chi(b)$$

and we have defined the product of two characters by

(7) $$\chi\chi'(a) = \chi(a)\chi'(a).$$

If two characters χ and χ' have the property

$$\chi(a) = \chi'(a) \quad \text{for all } a,$$

it follows that $\chi = \chi'$, and if two group elements a and a' have the property

$$\chi(a) = \chi(a') \quad \text{for all } \chi,$$

it follows that $a = a'$.

From these properties one sees that the roles of the characters χ and the group elements a can be interchanged. If Γ is the character group of G, then G can be regarded as the character group of Γ. The groups G and Γ form a "group pair" in the sense of Pontryagin. See L. Pontryagin: The Theory of Topological Commutative Groups, Annals of Math. 35, p. 389–395 (1934).

This duality can be extended to abelian topological groups. Besides the paper of Pontryagin just mentioned, the most important papers on this subject are:

A. Haar: Über unendliche kontinuierliche Gruppen, Math. Zeitschrift 33, p. 129–159 (1931),

J. von Neumann: Almost periodic Functions in a Group I, Transactions Amer. Math. Soc. 36, p. 445–492 (1934),

J.F. Alexander: On the Characters of Discrete Abelian Groups, Annals of Math. 35, p. 389–395 (1934),

E.R. van Kampen: Locally Bicompact Abelian Groups and their Character Groups, Annals of Math. 36, p. 448–463 (1935).

To fix the ideas, let me start with an example. The rotations about a point 0 in a plane form a compact abelian topological group. If x is the angle of a rotation, the group multiplication is defined by an addition of the angles modulo 2π.

As *characters* of an abelian topological group, one defines all *continuous bounded functions* χ satisfying (6). In our case, the only characters are

$$(8) \qquad\qquad \chi_n(x) = e^{inx}.$$

To every integer n corresponds just one character. The product of two characters corresponds to the sum of the integers:

$$\chi_m \chi_n = \chi_{m+n},$$

so the character group Γ is the additive group of the integers: a discrete abelian group. Conversely, the character group of Γ is G.

In Dirichlet's theory of Fourier series it is shown that the characters χ_n form a *complete orthogonal* set of periodic functions of x. The orthogonality relations are

$$(9) \qquad\qquad \int_0^{2\pi} \overline{\chi_m(x)}\, \chi_n(x)\, dx = \begin{cases} 2\pi & \text{if } m=n \\ 0 & \text{if } m \neq n, \end{cases}$$

and every continuous periodic function can uniformly be approximated by a finite sum of characters χ_n.

In this example we have seen that the character group of a compact topological group is a discrete group, and conversely. As we shall see presently, this is a general property of separable topological groups. If G is compact and separable, the character group Γ is discrete and countable, and conversely. This was proved by Pontryagin in his paper just quoted.

John von Neumann's paper starts with a general theory of almost periodic functions on an arbitrary group G, which may or may not be a topological group. A complex-valued function $f(x)$ defined on a group G is called *almost-periodic*, if every sequence of functions $f(a_\nu x b_\nu)$ contains a uniformly convergent subsequence. If G is a topological group, the functions $f(x)$ are required to be continuous. According to S. Bochner (Math. Annalen 96, p. 119–147, 1934) this general definition of almost-periodicity is equivalent to Harald Bohr's original definition, if G is the additive group of real numbers.

Von Neumann now defines mean values and scalar products of almost-periodic functions. He shows that the matrix elements of bounded representations are almost-periodic and satisfy certain orthogonality relations. Following Weyl and Peter, von Neumann uses an integral equation to prove that the matrix elements $d_{ik}(x)$ of irreducible bounded representations form a complete set of functions in the space of almost-periodic functions. For a summary of von Neumann's paper with simplified proofs see my "Gruppen von linearen Transformationen", p. 57–62.

In Part V of his paper, von Neumann considers the special case of locally compact, separable abelian groups. In this case the irreducible representations are all of degree 1, and their matrix elements $d_{ii}(x)$ are just the characters $\chi(x)$. It follows that these characters form a complete set of orthogonal functions in the space of almost periodic functions. Using a method of Haar, von Neumann shows that there exist "sufficiently many" characters to separate all group elements. This means: if $\chi(a) = \chi(b)$ for all characters χ, then $a = b$. Hence, the given group G and the group of characters Γ form a "group pair" in the sense of Pontryagin.

E.R. van Kampen extended the theories of Pontryagin and von Neumann to locally bicompact groups.

Part B
Characters of Finite Groups

We now turn to non-abelian finite groups.

The origins of the theory of group characters and representations of finite groups have been investigated very carefully by Thomas Hawkins in three papers:

1) The Origins of the Theory of Group Characters, Archive for History of Exact Sciences 7, p. 142–170 (1971),

2) Hypercomplex Numbers, Lie Groups, and the Creation of Group Representation Theory, same Archive 8, p. 243–287 (1972),

3) New Light on Frobenius' Creation of the Theory of Group Characters, same Archive 12, p. 217–243 (1974).

I have made a grateful use of these three papers.

The creator of the theory of characters of non-abelian finite groups is Georg Frobenius. As Hawkins has shown, Richard Dedekind played a vital role in the events that led to this creation. I shall now describe these events.

Dedekind's Introduction of the Group Determinant

Hawkins has drawn the attention to an unpublished manuscript of Dedekind written in 1886 and entitled "Gruppen-Determinante und ihre Zerlegung in wirkliche und überkomplexe Faktoren" (University Library Göttingen, Dedekind Manuscript V 5). In this manuscript, Dedekind associates with every element g_i of a finite group a variable x_i. The product $g_i g_j$ is again a group element, so it is associated with a variable x_k, which Dedekind calls x_{ij}.

Dedekind now forms the determinant of the x_{ij}. Each row and each column of this determinant is a permutation of the variables $x_1, \ldots, x_n$. Let Θ be this determinant. Later on in the same paper Dedekind changed his definition of the x_{ij}: he defined x_{ij} to be the variable associated with $g_i g_j^{-1}$. In what follows I shall adopt this latter definition. Putting

$$g_i g_j^{-1} = g_k$$

we now have

$$g_i = g_k g_j.$$

Let R be the group algebra generated by the group elements g_k and let

$$g_x = \Sigma g_k x_k$$

be a generic element of this algebra. Multiplying g_x with the group elements g_j, one obtains

$$g_x g_j = \Sigma (g_k g_j) x_k = \Sigma g_i x_{ij}.$$

So the matrix (x_{ij}) is just the matrix representing g_x in the regular representation of the group algebra, and Dedekind's "group determinant" Θ is the determinant of this matrix.

One of the first discoveries of Dedekind was: If G is abelian, the determinant Θ factorizes into linear factors with characters as coefficients:

(1)
$$\Theta = \prod_s \left[\sum_1^n \chi^{(s)}(g_i) x_i \right].$$

In February 1886, Dedekind decided to compute the group determinant for some non-abelian groups. He first considered the symmetric group S_3. The group elements are

$$g_1 = 1 \qquad g_4 = (2\ 3)$$
$$g_2 = (1\ 2\ 3) \quad g_5 = (1\ 3)$$
$$g_3 = (1\ 3\ 2) \quad g_6 = (1\ 2).$$

Dedekind found the decomposition

(2)
$$\Theta(x_1, \ldots, x_6) = (u+v)(u-v)(u_1 u_2 - v_1 v_2)^2$$

with

$$u = x_1 + x_2 + x_3$$
$$v = x_4 + x_5 + x_6$$
$$u_1 = x_1 + \rho x_2 + \rho^2 x_3$$
$$u_2 = x_1 + \rho^2 x_2 + \rho x_3$$
$$v_1 = x_4 + \rho x_5 + \rho^2 x_6$$
$$v_2 = x_4 + \rho^2 x_5 + \rho x_6$$

where ρ is a primitive cube root of unity.

Dedekind succeeded in factorizing the quadratic factor $u_1 u_2 - v_1 v_2$ by enlarging the field of coefficients $\mathbb{C}$ to an algebra of dimension 4 over $\mathbb{C}$. For the definition of this algebra, which is in fact isomorphic to the full matrix algebra of 2×2 matrices, see the paper 1) of Hawkins.

Dedekind next considered the dihedral group D_5 of order 10. Once more, he was able to factorize Θ into linear and quadratic factors.

Next he took the quaternion group. He obtained a factorization

$$(3) \qquad \Theta = (u_1 + u_2 + u_3 + u_4)(u_1 + u_2 - u_3 - u_4)(u_1 - u_2 + u_3 - u_4)$$
$$\cdot (u_1 - u_2 - u_3 + u_4)(v_1^2 + v_2^2 + v_3^2 + v_4^2)^2.$$

In this case, the quadratic factor represents the norm of a quaternion, so it can be factorized by introducing quaternions as coefficients:

$$v_1^2 + v_2^2 + v_3^2 + v_4^2 = (v_1 + i v_2 + j v_3 + k v_4)(v_1 - i v_2 - j v_3 - k v_4).$$

Dedekind did not publish his results, but in February 1895 he wrote a letter to Frobenius in which he raised the following cryptic question:

Do hypercomplex numbers with non-commutative multiplication also intrude in your research? But I do not wish to bother you with the request for an answer, which I will best obtain through your work (Dedekind's Werke, Vol. 2, p. 419-420).

In March 1896 Dedekind returned to the subject, defined the group determinant, stated his results concerning its factorization in the abelian case, and indicated what may happen when hypercomplex numbers are allowed as coefficients. In his next letter (April 3, 1896) he conjectured that the number of linear factors of Θ is equal to the order of the abelian group G/G', where G' denotes the commutator group, and that the linear factors of the group determinant correspond "in a certain way" to the characters of the group G/G'. He continued:

I would be delighted if you wished to involve yourself with these matters, because I distinctly feel that I will not achieve anything here.

Frobenius on the Group Determinant

Frobenius went to work at once. On April 12 he dispatched a long letter to Dedekind in which several highly interesting results concerning the prime factors of the group determinant are explained with full proofs. By the kindness of Clark Kimberling I have received copies of the correspondence between Dedekind and Frobenius. I shall now give a brief summary of the contents of this letter. For more details I may refer to pages 224–230 of Hawkins' paper 3).

After a few remarks on Hamiltonian groups, Frobenius proves that the number of linear factors in Θ is equal to the order of the abelian group G/G', as Dedekind had conjectured, and that the linear factors can be written as

$$F(x) = \sum_R \chi(R) x_R,$$

where x_R is the variable x_i associated with the group element R, while χ is a character of G in the classical sense:

$$\chi(RS) = \chi(R)\,\chi(S).$$

Frobenius notes that these characters χ are at the same time characters of the abelian group G/G' and conversely, which implies that their number is equal to the order of G/G'.

Frobenius' proof is essentially the one he published in his paper "Über Gruppencharaktere", which was presented to the Berlin academy on July 30, 1896.

In the letter, Frobenius next passes to the non-linear factors of Θ. I shall use Hawkins' notation (which is also Frobenius' notation in his later letters):

(4) $$\Theta = \Pi\Phi^e,$$

and I shall denote any prime factor of Θ by Φ, its degree by f, and its exponent in (4) by e.

Frobenius first derives an important relation, valid for every prime factor Φ:

(5) $$\sum_R \frac{\partial\Phi}{\partial x_R} x_{AR} = \Psi(A^{-1}).$$

As Hawkins notes, one can hardly overemphasize the importance of the relation (5). The complex function Ψ on the right hand side of (5) is just what Frobenius later called a character of G.

Frobenius' derivation of the relation (5) is very interesting. It has been explained by Th. Hawkins in his paper "The Creation of the Theory of Group Characters", Rice University Studies 64, p. 57–71 (1978). Frobenius defines his function $\Psi(A)$ for $A \neq E$ by

(6 a) $$\Phi = x_E^f + x_E^{f-1}\left(\sum_{A \neq E} \Psi(A)\,x_A\right) + \ldots$$

and for $A = E$ by

(6 b) $$\Psi(E) = f.$$

Using our hindsight, we can easily see that the function $\Psi(A)$, thus defined, is just the trace of A in an irreducible representation of the group G. We know that Φ is the determinant of the matrix X representing the generic element $\Sigma A x_A$ in an irreducible representation of the group algebra. Now if we replace the variable x_E by $x_E - \lambda$, we obtain from (6 a) and (6 b)

$$\det(X - \lambda I) = (-\lambda)^f + (-\lambda)^{f-1}\sum_A \Psi(A)\,x_A + \ldots$$

which implies that $\Sigma \Psi(A) x_A$ is the trace of the matrix X, and hence that $\Psi(A)$ is the trace of the matrix representing A in the irreducible representation corresponding to the prime factor Φ.

Frobenius next notes that the prime factors Φ have the multiplicative property

(7) $$\Phi(z) = \Phi(x)\,\Phi(y)$$

where

(8) $$z_A = \sum_{BC=A} x_B\, y_C.$$

From the modern point of view, formula (8) is just the multiplication formula for elements of the group algebra. In fact, if

$$x = \Sigma B x_B$$

and

$$y = \Sigma C y_C$$

are two elements of the group algebra, their product is just

$$z = \Sigma A z_A,$$

where z_A is defined by (8).

The notion "group algebra" was known to Frobenius. As we have seen in Chapter 6, the group algebra of a finite group had been defined by Cayley as early as 1854. In a letter to Dedekind, dated July 15, 1896, Frobenius explains:

Many times I have thought about how, to the multiplication of elements, one could adjoin another operation (addition).... Thus out of the elements $A, B, C, \ldots$ of the group, you form, by regarding them as hypercomplex units, linear combinations $xA + yB + zC + \ldots$ with scalar factors $x, y, z, \ldots$ (Hawkins' translation).

Now let us return to the letter of April 12. On page 12 of this letter, Frobenius calculates the decomposition of the group determinant Θ into prime factors in the case of the dihedral group of order 8 defined by the relations

$$A^4 = E, \quad B^2 = E, \quad B^{-1}AB = A^{-1}.$$

He finds a decomposition of the form

$$\Theta = (u+v)(u-v)(u_2+v_2)(v_2-u_2)(u_1 u_3 - v_1 v_3)^2.$$

In this example and in all examples computed by Dedekind the following theorems hold, which Hawkins calls Theorems A and B:

A. A linear change of variables is possible such that each factor Φ becomes a function of its own set of v variables.

B. $v = ef$.

On page 13 of his letter Frobenius raises the question whether A and B are generally true. The remainder of the letter is largely devoted to the proof of A and B.

First, Frobenius derives from (6) the "orthogonality relations"

(10)
$$\frac{h}{e} \Psi(AB^{-1}) = \sum_R \Psi(AR^{-1}) \cdot \Psi(RB^{-1})$$

(11)
$$0 = \sum_R \Psi'(AR^{-1}) \cdot \Psi(RB^{-1})$$

in which Ψ' is the function Ψ belonging to another prime factor Φ' different from Φ.

In the course of the derivation of (10) Frobenius notes that

(12)
$$\Psi(AB) = \Psi(BA),$$

and in the margin he notes the equivalent formula

(13)
$$\Psi(R^{-1}AR) = \Psi(A).$$

After having proved Theorem A, Frobenius announces: "Finally I believe to have found a proof of $v = ef$. But it does not please me at all and it must be possible to simplify it greatly." Indeed the proof is long and tedious.

At the end of this letter Frobenius asks Dedekind for help:

After your investigations on numbers from several units (that is, on algebras) you are certainly completely familiar with the methods of investigation, and you can provide simplifications. For my reasoning is so complicated that I myself do not rightly know where the main point* of the proof is…

and he adds a footnote:

* I believe it is contained in the equation $\Psi(AB) = \Psi(BA)$, for this implies that the system $(x_{AB^{-1}})$ is permutable with $\Psi(BA^{-1})$.

Five days later, on April 17, Frobenius writes jubilantly.

At the end of my last letter I gave up the search and requested your assistance. The next day I saw, if not the entire solution, at least the way to do it. My feeling that the equation $\Psi(AB) = \Psi(BA)$ provided the key did not deceive me. I still have a long way to go but I am certain I have chosen the right path…. Do you know of a good name for the function Φ…? Or should Ψ be called the character of Φ (which agrees for linear Φ)? (Hawkins' translation).

In the letter of April 17, Frobenius still hesitates whether he might call Ψ a character, but in later letters and in his paper of July 20 entitled "Ueber Gruppencharaktere" (Sitzungsberichte Akad. Berlin 1896, p. 985–1012) the functions Ψ are called "Gruppencharaktere".

In the letters as well in the paper on group characters, frequent use is made of a theorem concerning the characteristic roots of a set of commuting matrices. This theorem was proved in a paper "Ueber vertauschbare Matrizen", which was published in the Sitzungsberichte for 1896 on pages 601–614. I shall now summarize this paper, and next discuss the letters of April 17 and April 26 and the paper on group characters.

Frobenius on Commuting Matrices

The main theorem of Frobenius' paper "Über vertauschbare Matrizen" says:

III. Given m commuting $n \times n$-matrices $A, B, \ldots$, there exists an ordering $a_1, a_2, \ldots, a_n$; $b_1, b_2, \ldots, b_n; \ldots$ of the characteristic roots of $A, B, \ldots$ such that for any polynomial $f(u, v, \ldots)$ in m variables $u, v, \ldots$ the characteristic roots of $f(A, B, \ldots)$ are

$$f(a_1, b_1, \ldots), f(a_2, b_2, \ldots), \ldots, f(a_n, b_n, \ldots).$$

Special cases of this theorem are already present in an earlier paper of Frobenius "Über lineare Substitutionen und bilineare Formen", Crelle's Journal für Math. 84, p. 1–63 (1978), and Frobenius himself claims that he was already in possession of Theorem III when he wrote this earlier paper. Theorem III is also mentioned in an "Excurs" inserted in the letter of April 17, 1896.

If the m matrices are called $A_1, \ldots, A_m$, and if the polynomial f is a linear form in $u_1, \ldots, u_m$ with variable coefficients $x_1, \ldots, x_m$:

$$f(u_1, \ldots, u_m) = u_1 x_1 + \ldots + u_m x_m$$

it follows that the characteristic roots of

$$(14) \qquad\qquad A = A_1 x_1 + \ldots + A_m x_m$$

are the n linear functions of the x's

$$(15) \qquad\qquad r_1^{(s)} x_1 + \ldots + r_m^{(s)} x_m \qquad (s = 1, 2, \ldots, n)$$

where $r_i^{(s)}$ denotes the s-th characteristic root of A_i in the ordering of Theorem III. Since the determinant of A is the product of the characteristic roots, we have

$$(16) \qquad\qquad \det(A) = \prod_s \left(\sum_i r_i^{(s)} x_i \right).$$

In the continuation of his paper, Frobenius assumes $m = n$, and he takes for $A_1, \ldots, A_n$ the matrices of the regular representation of a commutative algebra. Let $e_1, \ldots, e_n$ be the basis elements of such an algebra, and let

$$(17) \qquad\qquad e_j e_k = \Sigma e_i a_{ijk}.$$

The matrices of the regular representation of this algebra are the matrices A_j with elements $a_{ik}^{(j)} = a_{ijk}$, and (17) implies

$$(18) \qquad\qquad A_j A_k = \Sigma A_i a_{ijk}.$$

Since the algebra is commutative, we have

$$A_j A_k = A_k A_j.$$

According to Theorem III, the s-th characteristic root of the left-hand side of (18) is $r_j^{(s)} r_k^{(s)}$, and of the right-hand side $\Sigma r_i^{(s)} a_{ijk}$, so (18) implies

(19) $$r_j^{(s)} r_k^{(s)} = \Sigma r_i^{(s)} a_{ijk},$$

which means that for every fixed s the numbers $r_1^{(s)}, \dots, r_n^{(s)}$ obey the same multiplication rules as the basis elements $e_1, \dots, e_n$ of the algebra. In other words, the mapping $\varphi^{(s)}$:

(20) $$\Sigma e_i x_i \to \Sigma r_i^{(s)} x_i$$

is a homomorphism of the algebra:

$$\varphi^{(s)}(a+b) = \varphi^{(s)}(a) + \varphi^{(s)}(b),$$
$$\varphi^{(s)}(a \cdot b) = \varphi^{(s)}(a) \cdot \varphi^{(s)}(b).$$

In Dedekind's paper of 1885 "Zur Theorie der aus n Haupteinheiten gebildeten komplexen Größen", which we have discussed in Chapter 6, a matrix (p_{ij}) had been defined:

(21) $$p_{ij} = \sum_{r,s} a_{rsr} a_{sij},$$

and Dedekind had proved: if $\det(p_{ij}) \neq 0$ the algebra is a direct sum of copies of $\mathbb{C}$. Frobenius now uses this result and proves: If $\det(p_{ij}) \neq 0$, the determinant of the $r_k^{(s)}$ is not zero, or in other words: The linear forms $\Sigma r_i^{(s)} x_i$ are linearly independent.

Using our hindsight, we may say that Frobenius developed a theory of irreducible representations of commutative algebras. If the ground field is algebraically closed, all irreducible representations of a commutative algebra are of degree one. The commuting matrices E_k of any representation of the algebra always have a common eigenvector, which generates a one-dimensional invariant subspace of the vector space on which the matrices operate. Modulo this subspace there is again an eigenvector, and so on. Thus the matrices can be written in triangular form, the elements below the main diagonal being zero. The diagonal elements $\varphi^{(s)}(a)$ define irreducible representations of the algebra.

If $\det(p_{ij}) \neq 0$, the algebra is what we call semi-simple, and according to Dedekind it is a direct sum of simple subalgebras which are copies of the complex number field $\mathbb{C}$. If this is assumed, the unit element of the algebra is the sum of the unit elements of the n simple algebras:

$$e = e_1 + e_2 + \dots + e_n$$

and every element of the algebra can be written as a sum

$$a = ae_1 + ae_2 + \ldots + ae_n.$$

The homomorphisms $\varphi^{(s)}$ can now be obtained as

$$a \mapsto ae_s$$

and they can be considered as representations of the algebra by complex numbers. It is clear the these n homomorphisms are linearly independent.

The Letter of April 17, 1896

Right at the beginning of this letter Frobenius expresses his feeling that the relation

(22) $$\Psi(AB) = \Psi(BA)$$

which he had derived in his earlier letter, provides the key to the entire solution of his problem. We shall now see how Frobenius used this key.

From (22) one derives, substituting $R^{-1}S$ and R for A and B,

(23) $$\Psi(R^{-1}SR) = \Psi(S)$$

which means that the function Ψ depends only on the class of conjugate elements to which S belongs. So each character $\Psi^{(s)}$ can be regarded as a vector having k components

$$\Psi_1^{(s)}, \ldots, \Psi_k^{(s)}$$

where k is the number of classes. The orthogonality relations (10) and (11) imply that the l vectors $\Psi^{(s)}$ are linearly independent, hence their number is at most equal to the number of classes:

$$l \leq k.$$

On page 3 of his letter, Frobenius introduces a set of integers $h_{\alpha\beta\gamma}$. One forms the products ABC, where A is in C_α, B in C_β, and C in C_γ, and one counts how many products ABC are equal to the unit element E. This number is $h_{\alpha\beta\gamma}$. It is symmetric in α, β, γ, and it is divisible by h_α.

Frobenius next investigates the question: What happens to the prime factors Φ if the variables x_R are restricted by the condition

(24) $$x_{AB} = x_{BA} \quad \text{or} \quad x_{R^{-1}SR} = x_S?$$

Following Hawkins, I shall call the restricted function Φ^*. On page 8 of his letter, Frobenius proves that Φ^* is a power of a linear function:

$$(25) \qquad \Phi^* = \frac{1}{f} \sum_R \{\Psi(R^{-1})x_R\}^f.$$

The conditions (24) can be interpreted as follows. The elements of the group algebra are sums

$$(26) \qquad x = \Sigma S x_S.$$

Now the conditions (24) imply a restriction to a subalgebra consisting of those sums x in which all elements S belonging to any class C_α have one and the same coefficient x_S. For the sake of clarity let us call this coefficient t_α. Then the sum (26) can be written as

$$(27) \qquad x = \Sigma e_\alpha t_\alpha$$

where e_α is the sum of all elements S of the class C_α.

The sums (27) form a subalgebra Σ of the group algebra. Its basis elements are the elements e_α, and its multiplication rules are

$$(28) \qquad e_\beta e_\gamma = \Sigma e_\alpha \frac{h_{\alpha'\beta\gamma}}{h_\alpha}.$$

Since $h_{\alpha'\beta\gamma} = h_{\alpha'\gamma\beta}$, the subalgebra Σ is commutative. In fact, it is the centre of the group algebra. Its composition constants are

$$(29) \qquad a_{\alpha\beta\gamma} = h_{\alpha'\beta\gamma}/h_\alpha.$$

These composition constants play a fundamental role in Frobenius' theory of group characters, as we shall see presently.

Frobenius next derives an extremely interesting formula

$$(30) \qquad f \sum_\gamma h_{\alpha\beta\gamma'} \Psi_\gamma = h_\alpha h_\beta \Psi_\alpha \Psi_\beta.$$

What does this formula mean? Let us rewrite it as

$$(31) \qquad \frac{h_\beta \Psi_\beta}{f} \frac{h_\gamma \Psi_\gamma}{f} = \Sigma \frac{h_{\alpha'\beta\gamma}}{h_\alpha} \frac{h_\alpha \Psi_\alpha}{f} = \Sigma a_{\alpha\beta\gamma} \frac{h_\alpha \Psi_\alpha}{f}.$$

This formula says that the complex numbers

$$(32) \qquad r_\alpha = \frac{h_\alpha \Psi_\alpha}{f}$$

obey the same multiplication rules as the basis elements e_α of the algebra Σ. In other words, the mapping

(33) $$\Sigma e_\alpha t_\alpha \mapsto \Sigma r_\alpha t_\alpha$$

is a homomorphic mapping of the algebra Σ into the complex number field.

The introduction of the numbers r_α defined by (32) is not my invention. The same numbers are denoted by $r_\alpha^{(s)}$ in Frobenius' paper "Über Gruppencharaktere", and the formula (31) occurs already in Frobenius' earlier paper "Über vertauschbare Matrizen" in the form

(34) $$r_j^{(s)} r_k^{(s)} = \Sigma r_i^{(s)} a_{ijk},$$

valid for any commutative algebra. See formula (19), which is identical with (34).

At the end of his letter Frobenius proves that $l = k$: the number of characters is equal to the number of classes in the group.

The Letter of April 26

At the beginning of this letter, Frobenius restates his earlier result: Every prime factor Φ of the group determinant can be transformed by a linear transformation of the variables x_k into a function of $v = ef$ variables, but not less. Next he reports that in all examples communicated by Dedekind $e = f$ holds. He says: "It would be wonderful if $e = f$", but he has been able to prove this only in the cases $f = 1$ and $f = 2$.

Next Frobenius introduces an algebra having composition constants $a_{\alpha\beta\gamma}$ defined by (29). He quotes Dedekind's paper of 1885 "Zur Theorie der aus n Haupteinheiten gebildeten complexen Zahlen". This paper, says Frobenius, "lies on my desk while I am writing this", and he adds "and there is no end to my astonishment". He shows that the algebra Σ satisfies Dedekind's condition

(35) $$\det(p_{\alpha\beta}) \neq 0.$$

This result is of fundamental importance for Frobenius. After having proved it, he says:

This simple proof cost me an incredible amount of strain and despair (Anstrengung und Verzweiflung).

The condition (35) implies, according to Dedekind, that the algebra Σ is a direct sum of copies of the complex field $\mathbb{C}$. Basing himself on this result, Frobenius is able to construct a complete set of representations of Σ by complex numbers

$$\Sigma e_\alpha t_\alpha \mapsto \Sigma r_\alpha^{(s)} t_\alpha$$

and to prove that the determinant of the $r_\alpha^{(s)}$ is not zero. Thus, a solid base for his theory of group characters was laid.

On page 10 of his letter, Frobenius returns to the factorization of the group determinant. He manages to compute the functions Ψ for five groups:

the tetrahedral group A_4 of order 12,
the icosahedral group A_5 of order 60,
the group $PSL(2, 7)$ of order 168,
the octahedral group S_4 of order 24,
the group S_5 of order 120.

By means of the equation

$$h = \Sigma\, ef$$

in which h is the order of the group, Frobenius calculates the products ef, and he finds that they always are squares.

For the tetrahedral group Frobenius actually calculated the factorization of the group determinant, and he found $e = f = 3$. In all cases, ef was just the square of $\Psi(E)$. He calls this "very curious, as long as it is not understood". At this stage, he was not yet able to prove

$$e = f = \Psi(E).$$

As early as 1893, Molien had proved that the group algebra of a finite group is a direct sum of full matrix algebras. I fully agree with Hawkins' remark:

Had he (Frobenius) known of Molien's work, he would certainly have seen its relevance, via the group algebra, to the study of the group determinant and to the matters that were currently puzzling him: the reasons why $e = f$ and $h = \Sigma f^2$ (Archive for History of Exact Sciences 12, p. 240).

Frobenius' Paper "Über Gruppencharaktere"

In this paper, which was presented to the Berlin academy on July 16, 1896, the line of thought explained in the letters is reversed and abridged. Instead of first defining the functions Ψ and next introducing the r_α defined by (32), Frobenius first introduces the r_α, and next defines the "characters" χ by the formula

(36) $$r_\alpha = \frac{1}{f} h_\alpha \chi_\alpha,$$

in which f is an arbitrary factor, which may be specified at a later stage. If f is equated to the degree of an irreducible factor of the group determinant, the characters χ are just the functions Ψ introduced in the letter of April 12, but in his published paper Frobenius does not mention the group determinant.

Frobenius defines the integers $h_{\alpha\beta\gamma}$ as in his letters, and he sets, once more,

(37) $$a_{\alpha\beta\gamma} = h_{\alpha'\beta\gamma}/h_\alpha.$$

He notes that the $a_{\alpha\beta\gamma}$ are the multiplication constants of a commutative algebra, to which the theorems proved by Weierstrass in 1884 and by Dedekind in 1885 can be applied.

Frobenius next recalls the results of his paper on commuting matrices. Putting

(38) $$a_{\alpha\beta} = \Sigma a_{\alpha\beta\gamma} x_\gamma$$

he finds that the characteristic roots of the matrix $(a_{\alpha\beta})$ are linear functions

(39) $$r_0^{(s)} x_0 + r_1^{(s)} x_1 + \ldots + r_{k-1}^{(s)} x_{k-1}.$$

Here $x_0, x_1, \ldots, x_{k-1}$ are variables corresponding to the k classes C_α, the first variable x_0 corresponding to the unity element E. He shows that Dedekind's condition

$$\det(p_{\alpha\beta}) \neq 0$$

is satisfied, and that the set of equations

(40) $$h_\beta h_\gamma \chi_\beta \chi_\gamma = f \sum_\alpha h_{\alpha'\beta\gamma} \chi_\alpha$$

has k different solutions

$$\chi_\alpha = \chi_\alpha^{(s)}, \quad f = f_s,$$

such that the determinant of the $\chi_\alpha^{(s)}$ is different from zero.

In §5 of this paper Frobenius introduces a new notation. Instead of $\chi_\alpha^{(s)}$ or χ_α he writes $\chi(A)$, where A is any element of the class C_α. He derives several interesting properties of the functions χ, for instance

$$h\chi(A)\chi(B) = f \sum_R \chi(AR^{-1}BR).$$

In the case of an abelian group, this relation reduces to

$$\chi(A)\chi(B) = f\chi(AB),$$

so, if one chooses $f = 1$, the characters χ of an abelian group coincide with the classical characters defined by Dirichlet and Dedekind.

The main ideas of the paper under discussion are all contained in the letters, but the letters contain much more information concerning the factorization of the group determinant. On this subject Frobenius composed another paper, which I shall now discuss.

The Proof of e = f and the Factorization of the Group Determinant

Between September 4 and September 6 of the same year 1896 in which Frobenius started his studies on the group determinant, he finally succeeded in proving $e=f$. In a letter to Dedekind dated September 6, he describes this event as follows:

> I will ... attempt to gather together the entire theory of the group determinant ... out of my highly scattered and disorganized papers. To *some* extent, however, such disorder is useful. That is, after my return home, I could no longer find the proof that I wrote to you long ago: If $f = 2$, then $e = 2$ also. After much torment I arrived at a new form of this proof and recognized here the possibility of generalization which I had completely despaired of in connection with the first proof (translated by Hawkins).

Now that $e=f$ was established, Frobenius was able to publish his results concerning the factorization of the group determinant. His paper "Über die Primfaktoren der Gruppendeterminante" was presented to the Berlin academy on December 3 (Sitzungsberichte 1896, p. 1343–1382).

Later on, in 1897, Frobenius learned about the paper of Molien and recast his own results in terms of matrices. He showed that his characters Ψ are just the traces of irreducible representations. In the next chapter I shall discuss his 1897 paper "Ueber die Darstellung der endlichen Gruppen durch lineare Substitutionen".

Chapter 13
Representations of Finite Groups and Algebras

The theory of representations of finite groups by matrices with complex matrix elements was developed, nearly simultaneously, by three authors: Theodor Molien, Georg Frobenius, and William Burnside. Their earliest papers on this subject are:

Th. Molien: Eine Bemerkung zur Theorie der homogenen Substitutionsgruppen, Sitzungsberichte der Naturforscher-Gesellschaft Dorpat (= Yurev in Estonia) 11, p. 259–276, presented April 24, 1897,

Th. Molien: Über die Anzahl der Variablen einer irreduziblen Substitutionsgruppe, same volume, p. 277–288, presented September 25, 1897,

G. Frobenius: Über die Darstellung der endlichen Gruppen durch lineare Transformationen, Sitzungsber. preuss. Akad. Berlin 1897, presented November 18, 1897,

W. Burnside: On the Continuous Group that is Defined by Any Given Group of Finite Order, Proceedings London Math. Soc. 29, p. 207–225, read January 10, 1898,

W. Burnside: Same title (second paper), same volume, p. 546–565, read June 9, 1898.

Thomas Hawkins has carefully investigated the genesis of these papers, and in particular the question of their independence. See his paper "Hypercomplex Numbers, Lie Groups, and the Creation of Group Representation Theory", Archive for History of Exact Sciences 8, p. 243–287 (1972). One of his conclusions is that Molien and Frobenius were completely independent of each other. Regarding Burnside, Hawkins concludes that he was fully aware on the connection of his paper with that of Frobenius on the prime factors of the group determinant, and that he "relied heavily for inspiration on the results in Molien's thesis on the characterization of simple hypercomplex systems". On the other hand, Burnside did not know the papers of Molien and Frobenius just mentioned.

The main results of the three papers are essentially the same, namely:

1. The regular representation of a finite group G is completely reducible.

2. Every irreducible representation is equivalent to one of the components of the regular representation.

3. If an irreducible component has degree f, it occurs f times in the regular representation. It follows that the sum Σf^2 is equal to the order of the group G.

4. The number of inequivalent irreducible representations is equal to the number of classes in the group G.

Molien also proved:

5. Every representation of G is completely reducible.

Burnside proved another important theorem, which he formulates thus:

It is further shown that, if g can be represented as a group of linear substitutions performed on m symbols, so that to the operation S_k of g there corresponds the substitution

(1)
$$x'_s = \sum_{i=1}^{i=m} a_{sik} x_i \quad (s = 1, 2, \ldots, m),$$

then the equations

(2)
$$x'_s = \sum_{i,k} a_{sik} x_i y_k \quad (s, i = 1, 2, \ldots, m; \ k = 1, 2, \ldots, n)$$

define a finite continuous group which is the direct product of a number of groups each of which is simply isomorphic with a general linear homogeneous group.

The two formulae of Burnside may be interpreted thus. The first formula (1) means that every element S_k of the group g is represented by a matrix A_k having elements a_{sik}. In the second formula these matrices A_k are multiplied by factors y_k and added, so as to obtain a matrix

(3)
$$A_y = \Sigma A_k y_k$$

representing the element $\Sigma S_k y_k$ of the group algebra. Burnside restricts himself to non-singular matrices A_y, which form a group, and he asserts that this group is a direct product of groups, each isomorphic with a group $GL(f, \mathbb{C})$.

If one considers *all* matrices A_y, without restricting oneself to non-singular matrices, one obtains an algebra: a representation of the group algebra. Molien proved that this algebra is a direct sum of full matrix algebras. Burnside's result is an immediate consequence of this theorem of Molien.

The methods of the three authors are quite different. Molien's investigations are based on the structure theory of algebras developed in his Thesis of 1891. Frobenius' paper is based on his theory of group characters, and Burnside's paper is based on Cartan's structure theory of semisimple Lie groups.

For further details I may refer to the paper of Hawkins quoted before.

Heinrich Maschke

The scientific career of Heinrich Maschke was sketched by Hawkins in Section 5 of his paper as follows:

Although he obtained his doctorate at Göttingen in 1880, and hence before Klein became a professor there, Heinrich Maschke (1853–1908) was a student of Klein's insofar as his mathematical activity was concerned. After receiving his doctorate, Maschke taught in a Gymnasium in Berlin until he returned to Göttingen in 1886 on a leave of absence. Under Klein's influence, he soon became a contributor to the general research program of Klein's form-problem and published a number of papers dealing with the determination of the invariants for particular groups of linear transformations and also with the linear groups themselves. Realizing he had no future as a mathematician in a Gymnasium and little chance of obtaining a university position in Germany, he immigrated to the United States in 1891. In 1892 he joined his friend Oskar Bolza, another student of Klein's, and E.H. Moore to form the Mathematics Department of the recently-established University of Chicago.

As we have seen in Chapter 8, E.H. Moore proved in 1896 that every finite group of linear transformations with complex coefficients has an invariant positive Hermitean form. In the same year 1896, A. Loewy independently obtained the same result.

This theorem was used by Maschke in 1898 to prove that every finite group of linear transformations is completely reducible. The principle of this proof is well known: If the vector space on which the group operates has an invariant linear subspace, there is an orthogonal subspace which is also invariant, and the whole space is the direct sum of two invariant subspaces. This construction is repeated until one obtains a decomposition of the whole space into irreducible subspaces. See H. Maschke: Über den arithmetischen Charakter der Substitutionen endlicher Substitutionsgruppen, Math. Annalen 50, p. 492–498 (1898).

Issai Schur

A simplified exposition of the representation theory of finite groups was given by Issai Schur in a paper entitled "Neue Begründung der Theorie der Gruppencharaktere", Sitzungsberichte preuss. Akad. Berlin 1905, p. 406–432.

In the introduction to his paper, Schur explains his relation to his predecessors thus:

The present paper contains a wholly elementary introduction into the theory of group characters, which was established by Frobenius, and which can also be designated as the theory of representations of finite groups by linear homogeneous substitutions.

An elementary justification of this theory has recently been given by Mr. Burnside. However, Burnside makes use of a tool which is, in principle, alien to the subject, namely the notion of Hermitean forms. Therefore, I don't consider it superfluous to communicate a new presentation of Frobenius' theory, which operates with still simpler tools.

Starting with a matrix representation of a finite group H of order h:

$$(4) \qquad\qquad a \rightarrow A,$$

Schur defines a *group matrix* X by the formula

$$(5) \qquad\qquad X = \Sigma A x_a,$$

the x_a being independent variables. He now formulates a fundamental theorem, which is known in the modern literature as *Schur's Lemma*, although it occurs already in Burnside's theory, as Schur himself admits. The lemma reads:

I. *Let X and X' be two irreducible group matrices of degrees f and f'. If P is a constant matrix having f rows and f' columns, for which the equation*

$$(6) \qquad\qquad XP = PX'$$

holds, then either P is zero, or X and X' are equivalent, and P is a quadratic matrix of degree $f = f'$ having a non-vanishing determinant.

This lemma holds for an arbitrary ground field. If the groundfield is algebraically closed (the only case considered by Schur), the lemma implies:

II. *If X is an irreducible group matrix of degree f, every constant matrix P commuting with X must be a multiple of the unit matrix.*

Schur next derives orthogonality relations for the matrix elements of irreducible representations.

Theorem III of Schur says:

Every group matrix X of degree n and rank r is equivalent to a group matrix having the form

$$\begin{pmatrix} X_1 & & & & \\ & X_2 & & & \\ & & \ddots & & \\ & & & X_m & \\ & & & & N_{n-r} \end{pmatrix}$$

in which $X_1, \ldots, X_m$ *are irreducible group matrices, while* N_{n-r} *is the zero matrix of degree* $n-r$.

Apart from the appended zero matrix, this theorem had already been obtained by Molien and Maschke. However, Maschke had used Hermitean forms, which can be formed only if the ground field is $\mathbb{C}$, whereas Schur's proof is completely elementary and valid for every ground field, provided the characteristic of the ground field is not a divisor of the order of the group G.

Representations of the Symmetric Group

Many authors in the first half of our century have investigated the representations of special finite groups. The literature on this subject is too vast to be discussed here. For a general survey I may refer to my "Gruppen von linearen Transformationen" (Springer-Verlag 1935, reprinted by Chelsea 1948), p. 78–84. Here I shall restrict myself to the representations of the symmetric group, which play an important role in the theory of invariants as well as in Quantum Mechanics.

The story begins with a sequence of papers by A. Young entitled "Quantitative Substitutional Analysis" in Proceedings London Math. Soc., of which the first two (1900 and 1902) are most important. The complete list of these papers is:

  I. Proc. (1) 33, p. 97–146 (1900)
  II. Proc. (1) 34, p. 361–397 (1902)
  III. Proc. (2) 28, p. 255–292 (1928)
  IV. Proc. (2) 31, p. 253–272 (1930)
  V. Proc. (2) 31, p. 273–288 (1930)
  VI. Proc. (2) 34, p. 196–230 (1932)
  VII. Proc. (2) 36, p. 36 (1933)
  VIII. Proc. (2) 37, p. 441 (1934).

In these papers, Young analyses the structure of the group algebra of the symmetric group S_n. He proves that the unity element e of this group is a sum of idempotent elements e_i:

(7)
$$e = e_1 + \dots + e_k$$
$$e_i^2 = e_i$$
$$e_i e_j = 0 \quad \text{for } i \neq j.$$

The units e_i are obtained as follows. A *Young Tableau* is an arrangement of the integers $1, 2, \dots, n$ in rows of non-increasing lengths, as in the following example:

$$1 \quad 3 \quad 5$$
$$2 \quad 4$$

Let P be the sum of all permutations transforming the rows into themselves, and let N be the alternating sum of all permutations transforming the columns into themselves. If T is the sum of the $n!$ products PN obtained by permuting the numbers in the tableau, we have

$$T^2 = cT \quad (c \neq 0).$$

Now the element

$$e_\alpha = T/c$$

is one of the desired nilpotent elements of the group algebra.

In his second paper, Young has calculated the value of the constant c. Let $\alpha_1, \dots, \alpha_h$ be the lengths of the rows in the tableau defining e_i. Then

(8)
$$c = \prod_{r,s} (\alpha_r - \alpha_s - r + s)^2 / \prod_r (\alpha_r + h - r)^2.$$

The e_α are in the centre Z of the group algebra A, and their number is equal to the number of the classes in the group S_n and hence to the dimension of the centre Z. It follows that Z is a direct sum of fields $e_i F$ isomorphic to the ground field F, provided the characteristic of F is not a prime factor of $n!$.

The decomposition (7) is of great importance in the theory of invariants. Let

$$f(u, v, \dots, w)$$

be a multilinear form in n rows of variables

$$u_1, \dots, u_m$$
$$v_1, \dots, v_m$$
$$\dots$$
$$w_1, \dots, w_m.$$

Now, if the vectors $u, \ldots, w$ are numbered from 1 to n, one can apply all elements of the group algebra of the symmetric group S_n as operators, operating on the form f, and one finds

$$(9) \qquad\qquad f = ef = e_1 f + \ldots + e_k f.$$

Whereas the multilinear form f is quite arbitrary, the terms $e_i f$ have strong symmetry properties. One term is symmetric in all variables, another term antisymmetric. On the importance of the expansion (9) for the theory of invariants see J.H. Grace and A. Young: The Algebra of Invariants (Cambridge 1903, reprinted by Chelsea), p. 359–364.

In the same year 1900, in which Young's first paper was published, Frobenius presented to the Berlin academy a paper "Über die Charaktere der symmetrischen Gruppe" (Sitzungsber. preuss. Akad. 1900, p. 516–534), in which the characters of S_n are determined. Frobenius first shows that to every partition

$$n = \alpha_1 + \ldots + \alpha_h \qquad (\alpha_1 \geq \alpha_2 \ldots \geq \alpha_h)$$

corresponds just one character. Frobenius puts

$$\beta_r = \alpha_r + h - r$$

and he proves that the character $\chi_\alpha(s)$ of a permutation s which is a product of cycles of lengths $\gamma_1, \gamma_2, \ldots$ is equal to the coefficient of

$$x_1^{\gamma_1} x_2^{\gamma_2} \ldots x_h^{\gamma_h}$$

in the polynomial

$$\prod_{q < r} (x_q - x_r) \cdot (x_1^{\gamma_1} + \ldots + x_h^{\gamma_1})(x_1^{\gamma_2} + \ldots + x_h^{\gamma_2}) \ldots .$$

The paper of Frobenius just cited is independent of Young's investigations, but in 1903 Frobenius used Young's Substitutional Analysis to determine the irreducible representations of S_n (see G. Frobenius: Über die charakteristischen Einheiten der symmetrischen Gruppen, Sitzungsber. preuss. Akad. Berlin 1903, p. 328–358). His main result may be expressed thus: Every product PN defined by a Young Tableau generates a minimal left ideal in the algebra A, and if the elements of this left ideal are multiplied by the group elements, one obtains an irreducible representation of the group. In this way, all irreducible representations of S_n can be obtained.

A simplified proof due to an oral communication of John von Neumann was presented in Volume 2 of my Algebra (first edition 1931, p. 203–207, English translation 1970, p. 93–97).

Other contributions to the representation theory of S_n are due to I. Schur, H. Weyl, A. Young, D.E. Littlewood and A.R. Richardson. See D.E. Littlewood: The Theory of Group Characters and Matrix Representations of Groups, second edition (1950), p. 59–146.

The Representation of Groups by Projective Transformations

In three papers in Crelle's Journal für Mathematik 127 (1904), 132 (1907), and 139 (1911), Issai Schur developed a method of constructing the representations of a finite group by projective transformations.

Let H be a finite group, and h its order. A projective representation of H is defined by a mapping

$$a \to A$$

having the property

$$ab \to \varepsilon_{a,b} AB.$$

The set of complex numbers $\varepsilon_{a,b}$ is called the *factor set* of the mapping. Two factor sets $\varepsilon_{a,b}$ and $\varepsilon'_{a,b}$ are called *associated* if

(10) $$\varepsilon'_{a,b} = (\gamma_a \gamma_b / \gamma_{ab}) \varepsilon_{a,b},$$

the γ_a being arbitrary factors $\neq 0$. The passage from one factor set to an associated set means a multiplication of the matrix A by a factor γ_a. Note that A and γA define the some projective transformation.

Every factor set is associated with a set consisting of h-th powers of unity. It follows that the classes of associated factor sets form a finite group M, the *multiplicator* of the group H. Let m be the order of M.

If G is a group containing in its centre a subgroup S such that

(11) $$G/S \cong H,$$

then every irreducible representation of G induces a projective representation of H, and all projective representations of H can thus be obtained.

If (11) holds and if S is contained in the commutator group of G and has the same order m as M, then all irreducible projective representations of H can be obtained from irreducible representations of G. If this is the case, G is called a *representation group* of H. A method to construct representation groups is presented in Schur's second paper in Crelle's Journal 132 (1907).

A group H is called *closed* if it is its own representation group, that is, if all projective representations are associated to ordinary matrix representations. For instance, all cyclic groups are closed. The quaternion group is closed. A group H is closed if all its Sylow groups are.

I. Schur has determined the representation groups of

$$SL(2, p^n), \quad PGL(2, p^n), \quad A_n, \quad \text{and} \quad S_n,$$

and R. Frucht has determined the representation groups of all finite abelian groups (Crelle's Journal 166, p. 16–29, 1931).

For more information about projective representations of finite groups I refer to my "Gruppen von linearen Transformationen", p. 84–88.

Emmy Noether

The first to develop a general representation theory of groups and algebras, valid for arbitrary ground fields, was Emmy Noether. In the Winter semester 1927–28 I attended her course of lectures "Hyperkomplexe Grössen und Darstellungstheorie" at the university of Göttingen, and she used my lecture notes as a basis for her publication "Hyperkomplexe Grössen und Darstellungstheorie" in Math. Annalen 30, p. 641–692. This publication has had a profound influence on the development of modern algebra. I shall now summarize its content.

In the introduction Emmy Noether states that in recent publications the structure theory of algebras and the representation theory of finite groups have been separated completely. She, on the other hand, aims at a purely arithmetical foundation, in which the structure theory and the representation theory of groups and algebras appear as a unified whole, namely as a theory of modules and ideals in rings satisfying finiteness conditions.

Emmy Noether's paper is divided into four chapters.

Chapter I. Group-Theoretical Foundations

§ 1. Groups with Operators. The notions in this section are due to Wolfgang Krull and Otto Schmidt. Whereas Krull restricted himself to abelian groups, the Russian group theorist Otto Schmidt introduced the general notion "Group with Operators", requiring only

$$\Theta(ab) = \Theta(a) \cdot \Theta(b).$$

See O. Schmidt: Über unendliche Gruppen mit endlicher Kette, Math. Zeitschrift 29, p. 34–41 (1928).

§ 2. The Isomorphy Theorems. In this section the well-known theorems concerning group homomorphisms, factor groups, and isomorphisms are generalized to groups with operators.

§ 3. Composition Series. In this section, Emmy Noether formulates the Jordan-Hölder theorem for groups with operators. For the proof she refers to page 57 of her paper "Abstrakter Aufbau der Idealtheorie in algebraischen Zahl- und Funktionenkörpern, Math. Annalen 96, p. 26–61 (1926).

§ 4. Direct Products and Intersections. Two equivalent definitions of the notion "direct product" are given. Every group satisfying the finiteness condition for descending chains of normal subgroups is a direct product of indecomposable factors. In his paper quoted before, Otto Schmidt has proved that such a decomposition is unique but for isomorphisms.

§ 5. Completely Reducible Groups. A group (with or without operators) is called completely reducible, if it is a direct product of simple groups. The factors are at the same time composition factors, and hence unique but for isomorphisms.

§ 6. *Moduli over a Field. Algebras.* If a module over a skew field K has a finite basis, it is a direct sum of one-dimensional modules:

$$G = a_1 K + \ldots + a_n K$$

and hence completely reducible. The number n is called the *rank* of G over K; today it is called the *dimension* of the vector space G.

§ 7. *Matrices.* A matrix over a skew field K having a right inverse also has a left inverse, and conversely. Such a matrix is called *regular*.

Chapter II. Non-Commutative Ideal Theory

§ 8. *The Homomorphy Theorem for Rings.* This theorem says: Every homomorphic image of a ring o is isomorphic to a residue class ring o/A, where A is a two-sided ideal in o. Today we call A the kernel of the homomorphism.

§ 9. *Idempotent Elements. Direct Sums of Right Ideals.* Let o be a ring with unity e. If o is a direct sum of right ideals:

$$o = R_1 + \ldots + R_n$$

and if

$$e = e_1 + \ldots + e_n \quad (e_i \text{ in } R_i)$$

then

$$e_i^2 = e_i, \quad e_i e_k = 0 \quad (i \neq k)$$

and

$$R_i = e_i o.$$

Conversely, if

$$e = \Sigma e_k$$

$$e_i^2 = e_i, \quad e_i e_k = 0 \quad (i \neq k)$$

and if one puts

$$R_i = e_i o$$

then o is a direct sum

$$o = R_1 + \ldots + R_n.$$

Hence a right decomposition

$$o = R_1 + \ldots + R_n = e_1 o + \ldots + e_n o$$

implies a left decomposition

$$o = L_1 + \ldots + L_n = o e_1 + \ldots + o e_n.$$

§ 10. Sums of Two-Sided Ideals. If o is a ring with unity, and if o is a sum of two-sided ideals

(12) $$o = A_1 + \ldots + A_n$$

then

$$A_i^2 = A_i, \quad A_i A_k = 0 \quad (i \neq k).$$

If the A_i cannot be further decomposed into two-sided ideals, the decomposition (12) is unique.

§ 11. The Centre. To every decomposition (12) of a ring o corresponds a decomposition of the centre Z:

$$Z = Z_1 + \ldots + Z_n$$
$$Z_i = A_i \cap Z.$$

§ 12. Nilpotent Ideals. An ideal C is nilpotent, if a power of C is zero.

If a ring o has a nilpotent right ideal R, it also has a nilpotent two-sided ideal. The sum of two nilpotent right ideals is a nilpotent right ideal. If the finiteness condition for increasing chains of right ideals is satisfied, there is a maximal nilpotent two-sided ideal, which contains all nilpotent right or left ideals. It is called the *radical* of o. If o is a ring without radical, so is the centre Z.

§ 13. Completely Reducible Rings. A ring is called *right completely reducible*, if it is a direct sum of simple right ideals.

A right completely reducible ring with unity has no radical.

Conversely: A ring without radical with descending chain condition for right ideals is right completely reducible, and it has a unity element.

§ 14. Completely Reducible Simple Rings. If a ring o with unity is simple and right completely reducible, all right ideals are isomorphic and the ring is a complete matrix ring over a skew field Λ.

Conversely: the ring of $n \times n$-matrices over Λ is simple and completely reducible.

Chapter III. Modules and Representations

§ 15. Representations and Representation Modules. Let o be a ring, and K a ring with unity. In the applications K is always a skew field.

A representation of degree n is a homomorphism

$$o \rightarrow \mathcal{O}$$

where $\mathcal{O}$ is a ring of $n \times n$-matrices over K.

A *representation module* is a "double module" M, which is a left o-module and a right K-module:

$$oM \subseteq M, \quad MK \subseteq M,$$

and which is a direct sum

$$M = x_1 K + \ldots + x_n K,$$

the unity element of K being a unity operator.

Every representation module generates a representation:

$$c x_k = \Sigma x_i \gamma_{ik}.$$

Conversely: Every representation is generated by a representation module.

Operatorisomorphic representation modules generate equivalent representations, and conversely.

§ 16. *Reducible Representations.* If K is a skew field, every submodule A of a representation module M has a basis $(y_1, \ldots, y_r)$ which can be completed to form a basis of M:

$$M = y_1 K + \ldots + y_r K + z_1 K + \ldots + z_{n-r} K.$$

If this is supposed, the matrices of the representation assume the form

$$C = \begin{pmatrix} R & 0 \\ S & T \end{pmatrix}$$

and the representation is called *reducible*.

A composition series

$$M \supset A_1 \supset \ldots \supset A_e = 0$$

leads to matrices of the form

$$C = \begin{pmatrix} R_{11} & & & \\ R_{12} & R_{22} & & \\ \vdots & \vdots & & \\ R_{1e} & R_{2e} & \ldots & R_{ee} \end{pmatrix}$$

in which the R_{ii} are irreducible representations, generated by the composition factors A_{i-1}/A_i. The theorem of Jordan-Hölder implies that the R_{ii} are unique but for equivalence and but for their order.

§ 17. *Direct Sum Decompositions of Representation Modules.* Let M be an o-module, and o a ring having a unity element e. Then every element m of M can be decomposed:

$$m = em + (m - em)$$

and thus M becomes a direct sum

$$M = M_e + M_o$$

such that e is a unity operator for M_e, and $eM_o=0$. Thus one can always restrict oneself to the study of modules for which e is a unity operator. In the case of a representation module this means that e is represented by the unity matrix E.

If o is a direct sum of two-sided ideals

$$o = A_1 + \ldots + A_s$$

then every o-module for which e is a unity operator is a direct sum

$$M = A_1 M + \ldots + A_s M.$$

Hence one can always restrict oneself to representations of *indecomposable* rings, that is, of rings which cannot be decomposed into two-sided ideals.

§ 18. *Modules and Representations of Completely Reducible Rings.* Of fundamental importance for Emmy Noether's representation theory is the theorem:

Let o be simple and completely reducible, and let M be a finite o-module for which the unity element of o is a unity operator. Then M is completely reducible, and the irreducible components of M are isomorphic to the simple left ideals L_i of o.

The proof is extremely simple. Let o and M be written as

$$o = L_1 + \ldots + L_n$$
$$M = (o m_1, \ldots, o m_k).$$

Then

$$M = (\ldots, L_i m_k, \ldots)$$

and if those modules $L_i m_k$ are left out of account, which are contained in the sum of the preceding ones, M becomes a direct sum of modules $L_i m_k$ isomorphic to L_i.

§ 19. *The Simple Composition Factors of Modules and Representation Modules.* Let o be a ring with maximal and minimal condition for left ideals, and let R be the radical. Then every simple o-module is either annihilated by o, or it is isomorphic to a simple left ideal in o/R. If, in particular, M is a module of finite rank over a field Γ, it is a representation module, and one can conclude that all irreducible representations (apart form the zero representation) are generated by the simple left ideals of o/R.

Chapter IV. Representations of Groups and Algebras

§ 20. *Algebras.* If o is an algebra over a field Γ, and if one restricts oneself to those left and right ideals that are Γ-modules, the finite chain conditions for these ideals are satisfied, and one can apply the theorems of § 19 to determine the irreducible representations. The result is: All irreducible representations are generated by the simple left ideals of

$$\bar{o} = o/R,$$

where R is the radical of o. Since $\bar{o}$ is semi-simple, it is a direct sum of simple algebras:

(13) $$\bar{o} = A_1 + \ldots + A_t$$

and there are just as many irreducible representations as there are terms in the sum (13).

In particular, if Γ is algebraically closed, every subalgebra A_i is a full matrix ring of rank n_i^2 over Γ. In this case every irreducible representation contains just n_i^2 linearly independent matrices, and every completely reducible representation contains

$$n_1^2 + \ldots + n_r^2$$

linearly independent matrices, where $n_1, \ldots, n_r$ are the degrees of the inequivalent components of the representation. This theorem is due to G. Frobenius and I. Schur: "Über die Äquivalenz der Gruppen linearer Substitutionen", Sitzungsber. preuss. Akad. Berlin 1906.

§ 21. *Extension of the Ground Field. The Representations of the Centre.* If $o = a_1 \Gamma + \ldots + a_n \Gamma$ is an algebra, and Ω an extension of the ground field Γ, one can form a new algebra

$$o\Omega = a_1 \Omega + \ldots + a_n \Omega$$

retaining the rules of multiplication for the basis elements a_i. Thus one obtains an algebra $o\Omega$ over Ω. If $o\Omega$ is a ring without radical, so is o, but the converse is not always true. The question whether $o\Omega$ has a radical depends on the structure of the centre Z of o.

Let Ω be the algebraically closed (algebraic) extension of Γ. If o is without radical, so is Z, and hence Z is a sum of fields

$$Z = Z_1 + \ldots + Z_t.$$

If one of these fields is an inseparable extension of Γ, then $Z\Omega$ has a radical, and so has $o\Omega$. On the other hand, if all fields Z_i are separable over Γ, the algebra $o\Omega$ has no radical, and all representations of o in Ω are completely reducible.

A representation of o is called *absolutely irreducible* if it remains irreducible after passing from Γ to the algebraically closed extension Ω. Emmy Noether now proves:

If $o\Omega$ has no radical, then in every absolutely irreducible representation the elements z of the centre Z are represented by diagonal matrices $E\zeta$. Every absolutely irreducible representation of o induces a linear representation $z \to \zeta$ of Z. The number of classes of absolutely irreducible representations equals the rank of Z.

§ 22. *Application to Abelian Groups.* The absolutely irreducible representations of abelian groups are just the well-known characters χ.

§ 23. *The Determinant of an Algebra.* By means of indeterminates $x_1, \ldots, x_n$ one can form the "generic element" of an algebra o:

$$w = a_1 x_1 + \ldots + a_n x_n.$$

If $a \to A$ is a representation, the generic element w is represented by

$$W = A_1 x_1 + \ldots + A_n x_n.$$

The determinant of the matrix W is called the *system determinant.* The regular representation gives rise to the *regular system determinant.* If the representation $a \to A$ is reducible, the matrix W can be written as

$$W = \begin{pmatrix} W_{11} & & & \\ W_{21} & W_{22} & & \\ \vdots & \vdots & & \\ W_{e1} & W_{e2} & \cdots & W_{ee} \end{pmatrix}$$

and $\mathrm{Det}(W)$ is the product of the system determinants of the irreducible representations $w \to W_{ii}$.

§ 24. *Traces and Characters.* For every representation $a \to A$ the trace of A is a linear function of a:

$$Tr(a + b) = Tr(a) + Tr(b)$$
$$Tr(c\alpha) = \alpha \, Tr(c).$$

The traces in the absolutely irreducible representation are called *characters* $\chi(a)$.

If o has no radical and if the characteristic of the field Γ is zero, every class of equivalent representations is uniquely determined by its traces.

§ 25. *Discriminants.* If $Tr(a)$ is the trace of a in the regular representation, one can form the determinant of the matrix formed by the traces $Tr(a_i a_k)$. This determinant is called the *discriminant.* If one passes to another basis, the discriminant is multiplied by the square of the transformation determinant.

The discriminant is zero if o has a nilpotent ideal. On the other hand, if o has no radical and if the characteristic of the field is zero, the discriminant is not zero.

Note that the matrix formed by the traces

$$p_{ik} = Tr(a_i a_k)$$

was used already by Dedekind and Molien as a criterion for semi-simplicity.

§ 26. *The Group Algebra.* If o is the group algebra of a finite group of order h, the discriminant of o is

$$D = h^h e.$$

Hence, if the order of the group is not divisible by the characteristic of the ground field, the group algebra o has no radical, all representations are completely reducible, and the number of absolutely irreducible representations is equal to the number of classes.

I have followed Emmy Noether's exposition as closely as possible in order to give the reader an idea of the personal style of her exposition.

Chapter 14
Representations of Lie Groups and Lie Algebras

According to Sophus Lie, every local representation of a Lie group G by linear transformations is generated by a representation of the Lie algebra L_G of G. The matrices of this representation are linear combinations

$$A = A_1 x_1 + \ldots + A_n x_n,$$

and a neighbourhood of the unity element of G is represented by the matrices

$$\exp A = \exp(A_1 x_1 + \ldots + A_n x_n),$$

the x_i varying in a neighbourhood of the origin in $\mathbb{R}^n$ or $\mathbb{C}^n$. It is true that Lie does not use the modern expressions "exp" and "neighbourhood", but his statements are equivalent to what I have just said.

The first to develop a general representation theory of Lie groups was Elie Cartan.

Cartan's Theory

According to Elie Cartan, if one wants to find all irreducible Lie groups of linear transformations, it is sufficient to find all irreducible representations of *simple* Lie groups. This was proved on p. 99 of Cartan's paper of 1909 "Les groupes de transformations continus, infinis, simples", Annales de l'Ecole normale (3) 26, p. 93–161.

If the ground field is the complex number field $\mathbb{C}$, there are four infinite sequences and five exceptional types of simple Lie algebras. For these types Cartan has determined all irreducible representations in his fundamental paper "Les groupes projectifs qui ne laissent invariante aucune multiplicité plane", Bulletin de la Société Math. de France 41, p. 53–96 (1913).

As we have seen in Chapter 9, the four infinite sequences of simple groups discovered by Sophus Lie are

    A: the projective linear groups $PGL(n, \mathbb{C})$ with $n > 1$,
    B: the projective orthogonal groups $PO(2n, \mathbb{C})$ with $2n > 4$,
    C: the projective symplectic groups $PSp(2n, \mathbb{C})$,
    D: the projective orthogonal groups $PO(2n-1, \mathbb{C})$ with $2n-1 > 1$.

Let us begin with type A. Instead of $PGL(n, \mathbb{C})$ we may also consider the locally isomorphic group $SL(n, \mathbb{C})$. Its elements are the $n \times n$ matrices having determinant 1. Its infinitesimal transformations are the matrices having trace zero:

(1) $$a_{11} + a_{22} + \ldots + a_{nn} = 0.$$

There are two methods to find the representations of this group SL: the global and the infinitesimal method. Both methods are purely algebraic, and both end up with the same results.

The Global Method

The global method starts with the group $GL(n, F)$, the field F being an arbitrary field of characteristic zero. It is a method which yields all those representations of the group GL, in which the elements of the matrix $T(A)$ representing a matrix A are polynomials in the matrix elements $a_{\kappa\lambda}$ of A.

The global method was inaugurated by Issai Schur in his doctoral thesis (Berlin 1901). The method was further developed by Hermann Weyl in 1925 and by D.E. Littlewood and A.R. Richardson in 1934. The most important papers on the subject are:

I. Schur: Über eine Klasse von Matrizen, die sich einer gegebenen Matrix zuordnen lassen, Diss. Berlin 1901 = Werke I, p. 1–70;

H. Weyl: Theorie der Darstellung kontinuierlicher halbeinfacher Gruppen durch lineare Transformationen I, Math. Zeitschrift 23, p. 271–309 (1925), or Selecta Hermann Weyl, p. 263–298;

I. Schur: Über die rationalen Darstellungen der allgemeinen linearen Gruppe, Sitzungsber. preuss. Akad. Berlin 1927, p. 59–75,

D.E. Littlewood and A.R. Richardson: Group Characters and Algebra, Philos. Transactions Roy. Soc. A 223, p. 99–141 (1934).

In explaining the global method, I shall follow the exposition in my "Gruppen von linearen Transformationen" (Springer-Verlag 1935), p. 88–91, which is largely inspired by the work of Hermann Weyl.

The central idea underlying the global theory is very simple. First one shows that every representation in which the matrix elements are polynomials in the $a_{\kappa\lambda}$ can be decomposed into representations in which the matrix elements are *homogeneous* polynomials. If m is their degree, we shall say that we have a representation of *height m* (in German: *Stufe m*).

A fairly general representation of height m is the *tensor representation*. If $u_1, \ldots, u_n$ are the basis vector of a vector space on which $GL(n, F)$ operates, and $v_1, \ldots, v_n$ the basis vectors of another copy of the same vector space, etc., then the products of m factors

$$u_\lambda v_\mu \ldots w_\nu$$

are the basis elements of a tensor space V of height m. The elements of this space are *tensors*

$$t = \Sigma t_{\lambda \mu \ldots \nu} \, u_\lambda \, v_\mu \ldots w_\nu.$$

The matrix elements of the tensor representation of $GL(n, F)$ are

$$a_{\kappa_1 \ldots \kappa_m, \, \lambda_1 \ldots \lambda_m} = a_{\kappa_1 \lambda_1} a_{\kappa_2 \lambda_2} \ldots a_{\kappa_m \lambda_m}.$$

The linear closure S of this set of matrices consists of all symmetrical matrices, that is, of those matrices, in which the matrix elements

$$a_{\kappa_1 \ldots \kappa_m, \, \lambda_1 \ldots \lambda_m}$$

remain unchanged if one and the same permutation Q is applied to the κ as well as to the λ. The proof, due to Hermann Weyl, is very simple.

Every permutation Q in S_m, if applied to the κ as well as to the λ, induces a linear transformation

$$\tau \to Q\tau$$

of the tensor space into itself. Let R be the set of these linear transformations. According to the theorem of Weyl just quoted, the linear closure S consists of those transformations of V into itself that are permutable with the matrices of R. But R is a representation of the symmetric group S_m, and hence completely reducible. By a theorem of Issai Schur, it is very easy to determine all linear transformations commuting with a given completely reducible set of matrices. It follows that S is a direct sum of full matrix rings over the ground field F, and that *all irreducible representations of GL of height m are contained in the tensor representation.*

We have seen that the irreducible representations of S_m are determined by Young tableaux like

$$\begin{array}{ccc} 1 & 3 & 5 \\ 2 & 4 & \end{array}$$

If P is the sum of the permutations within the rows of such a tableau, and N the alternating sum of the permutations within the columns, the product PN is (apart from a constant factor) an idempotent element of the group algebra of S_m. Now if the operator PN is applied to all tensors t, one obtains a linear set of tensors, which is transformed into itself by all transformations of GL, and these transformations form an irreducible representation of GL. All rational representations of height m can be obtained in this way. That's all!

The Infinitesimal Method

The infinitesimal method of Elie Cartan is based on the structure theory of simple Lie algebras. I shall follow the exposition of Weyl (Selecta Hermann

Weyl, p. 266–278), starting with the simplest case: the Lie algebra L_G of the special linear group $G = SL(n, \mathbb{C})$.

This group contains a maximal abelian subgroup H formed by the diagonal matrices of determinant 1. The infinitesimal transformations of this group are given by diagonal matrices of trace zero:

$$h(\lambda_1, \ldots, \lambda_n) = \begin{pmatrix} \lambda_1 & & & \\ & \lambda_2 & & \\ & & \ddots & \\ & & & \lambda_n \end{pmatrix},$$

$$\lambda_1 + \ldots + \lambda_n = 0.$$

Let e_{ik} $(i \neq k)$ be the matrix having 1 in row i and column k, and zero everywhere else. The e_{ik} form, together with the matrices $h(\lambda)$, a basis of the Lie algebra L. The commutator relations containing the matrices $h(\lambda)$ are:

(2)
$$[h, h'] = 0$$
$$[h, e_{ik}] = (\lambda_i - \lambda_k) e_{ik}.$$

Because of these relations, the linear forms $\lambda_i - \lambda_k$ are called the *roots* of the Lie algebra. The element $e_\alpha = e_{ik}$ is said to *belong* to the root $\alpha = \lambda_i - \lambda_k$.

Let h_α be the $h(\lambda)$, for which $\lambda_i = 1$ and $\lambda_k = -1$, all other λ's being zero. The commutators relations between the e_α are

$$[e_\alpha, e_{-\alpha}] = h_\alpha$$

$$[e_\alpha, e_\beta] = \begin{cases} \pm e_{\alpha+\beta} & \text{if } \alpha + \beta \text{ is a root} \\ 0 & \text{if not.} \end{cases}$$

Cartan's problem is to find a mapping

$$u \to U$$

such that

$$[u, v] \to [U, V] = UV - VU.$$

This means: the matrices $H(\lambda_1, \ldots, \lambda_n)$ and E_α have to satisfy the conditions

(3)
$$[H, H'] = 0, \qquad [H, E_\alpha] = \alpha E_\alpha,$$
$$[E_\alpha, E_{-\alpha}] = H_\alpha, \qquad [E_\alpha, E_\beta] = \pm E_{\alpha+\beta} \quad \text{or } 0.$$

The vectors on which the matrices U operate will be denoted by latin letters. If a vector satisfies the relation

$$He = \Lambda e$$

in which Λ is a linear form in the λ's, we shall say that e has *weight* Λ.

If a vector x has weight Λ, then Hx also has weight Λ, and $E_\alpha x$ has weight $\Lambda + \alpha$.

If $e \neq 0$ is a vector of weight Λ, one can form an invariant linear subspace spanned by the vectors

$$e, E_\alpha e, E_\alpha E_\beta e, \text{ etc.}$$

If the representation is irreducible, this subspace must be the whole vector space V, so V has a basis consisting of vectors each having a definite weight.

Let Λ be a weight:

(4)
$$\Lambda = m_1 \lambda_1 + \ldots + m_n \lambda_n,$$

and let Λ_α be the value of Λ for $H = H_\alpha$, that is,

$$\Lambda_\alpha = m_i - m_k.$$

The weight Λ is called *extreme* with respect to α, if $\Lambda + \alpha$ is no longer a weight. Starting with an extreme weight Λ, one can form the sequence

$$E_{-\alpha} e = e_1, \qquad E_{-\alpha} e_1 = e_2, \ldots .$$

In this sequence let e_{h+1} be the first zero vector. The vectors

$$e, e_1, \ldots, e_h$$

have weights

$$\Lambda, \Lambda - \alpha, \ldots, \Lambda - h\alpha.$$

One can also return from e_i to e_{i-1}:

$$E_\alpha e_i = \mu_i e_{i-1}.$$

The final result of Cartan's investigation can be formulated thus. Let

$$x_1, \ldots, x_n$$
$$y_1, \ldots, y_n$$
$$\ldots$$

be m sets of "cogredient variables". This means: is a linear transformation T is applied to the x, the same transformation will be applied to the y, the z, etc. Let

$$m = \alpha_1 + \alpha_2 + \ldots + \alpha_r$$

be a partition of the integer m subject to the conditions

$$\alpha_1 \geq \alpha_2 \geq \ldots \geq \alpha_r \geq 0 \quad \text{and} \quad r \leq n.$$

If one applies to the form

$$f = x_1^{\alpha_1 - \alpha_2}(x_1 y_2 - x_2 y_1)^{\alpha_2 - \alpha_3} \ldots (\Sigma \pm x_1 y_2 z_3 \ldots)^{\alpha_r}$$

all linear transformations T of the $x, y, \ldots$, one obtains a linear set of forms Tf. Applying to these forms the infinitesimal transformations of $SL(n, \mathbb{C})$, one obtains an irreducible representation of the Lie algebra L_G, and all irreducible representations of L_G can be obtained in this way.

Evidently, these infinitesimal representations can be extended to global representations of SL, and even to global representations of the general linear group GL. One has only to apply to the forms Tf all non-singular linear transformations S of GL. The representations of GL thus obtained are just the rational representations which Issai Schur had found in 1901. In a letter to Hermann Weyl, dated March 1st, 1925, Cartan states that he had determined the irreducible representations of the Lie algebra without knowing about the research of Issai Schur.

Cartan applied the same method to the Lie algebras of types B, C, and D. For the sympletic groups (type C) the result is similar: all irreducible infinitesimal representations can be extended to global representations which are contained in the tensor representations of $Sp(n, \mathbb{C})$.

For the orthogonal groups (types B and D) the situation is different. In Chapter 10, in the section "Spinors in n Dimensions", I have described two-valued "spinor representations" of the orthogonal groups, which are implicit already in the work of Lipschitz on sums of squares, and which have been established explicitly by Brauer and Weyl. Cartan, who systematically determined all infinitesimal irreducible representations by his method of weights, naturally also found these two-valued representations. See E. Cartan: Oeuvres complètes I, p. 387–393.

By the same method, Cartan also determined the infinitesimal irreducible representations of the five exceptional simple Lie algebras.

Hermann Weyl

The story of how Hermann Weyl arrived at his admirable representation theory of semi-simple groups is very curious. In April 1918, Weyl published the first edition of his famous book on general relativity: "Raum, Zeit, Materie". In the fourth edition of this book (1921) Weyl inserted a section (§ 18) entitled "Gruppentheoretische Auffassung der Raummetrik", which may be summarized thus.

In a Riemannian manifold we have at any point P a tangential vector space of dimension n. In this vector space a "Pythagorean metric" is defined by a quadratic form

$$ds^2 = g_{ik}\, dx_i\, dx_k.$$

The group of rotations transforming this form into itself is a Lie group G of dimension

$$\tfrac{1}{2}n(n-1).$$

Weyl now asks: What properties of the group G characterize the quadratic "Pythagorean metric"?

Weyl first requires the transformations of G to preserve volumes, that is, to have determinant one. His second condition demands the existence and unicity of *parallel displacements* of vectors to infinitely near points preserving the metric. This condition implies that the dimension of the group G is at most $\tfrac{1}{2}n(n-1)$. Weyl's third condition requires the existence of sufficiently many *congruent displacements* of vectors to infinitely near points. This condition implies that the dimension of G is at least $\tfrac{1}{2}n(n-1)$.

Weyl conjectures that the only groups satisfying these conditions are the rotation groups defined by non-singular quadratic forms. He states that he has proved this conjecture for $n=2$ and $n=3$.

Early in 1921, Weyl succeeded in proving his conjecture. His paper "Die Einzigartigkeit der Pythagoreischen Maßbestimmung" was submitted to the Math. Zeitschrift in April 1921 and published in Vol. 12 in January 1922.

Independent of Weyl, Elie Cartan found a simpler proof of Weyl's conjecture, based on his representation theory of Lie groups. His proof was announced in a Comptes-Rendus note in 1922 (C.R. Acad. Sc. Paris 175, p. 82), and published in 1923 in a paper entitled "Sur un théorème fondamental de M.H. Weyl", Journal de Math. 2, p. 167-192 (Oeuvres complètes III, p. 63-88).

Cartan sent an offprint of his C.R.-note to Weyl. An extremely interesting correspondence between Cartan and Weyl followed. The dates of the most important letters are:

from Weyl October 5, 1922,
from Cartan October 9, 1922,
from Cartan June 28, 1923,
from Cartan March 1st, 1925,
from Weyl March 22, 1925,
from Cartan March 28, 1925.

I am indebted to Dr. B. Glaus, librarian at the E.T.H. Zurich, for showing me the first three letters from Cartan to Weyl. Thereupon Elie Cartan's son Henri Cartan sent me copies of the other three letters. The letters and copies are now in the archive of the E.T.H. Zurich.

From the first letter, in which Weyl thanks Cartan for sending his C.R.-note, we see that Weyl was not familiar with Cartan's general theory of Lie groups before October 1922. He writes:

Unbewandert auf den gebahnten Strassen der allgemeinen Theorie der kontinuierlichen Gruppen, deren Anlage und Ausbau man Ihrer Meisterschaft verdankt, habe ich auf eigene Faust einen steilen, unbequemen Fusspfad durch das Gestrüpp zum Ziel gebahnt. Ich zweifle nicht daran, dass Ihre Methode der Natur der Sache besser entspricht...

I must confess that I am not able to translate the poetic imagery of Hermann Weyl into English. I am not a poet.

Cartan's answer is very courteous. He praises Weyl's "important theorem", and he adds: "Once the theorem being recognized as true, the simplicity of its proof is nothing compared with its philosophical significance ("sa portée philosophique").

In June 1923, Weyl sent Cartan a copy of his book "Mathematische Analyse des Raumproblems". Cartan answered at once: "Je vous remercie bien vivement d'avoir bien voulu m'envoyer votre beau livre."

From the correspondence we see that Weyl began to study Cartan's theory in October 1922, because he saw that this theory was a great help in his study of the "space problem". Very soon he had mastered Cartan's theory, and he even made important contributions himself. His investigations resulted in three beautiful papers, of which he sent copies to Cartan in February 1925. The papers were published in Math. Zeitschrift 23 (1925) and 24 (1926) under the general title "Theorie der Darstellung kontinuierlicher halb-einfacher Gruppen durch lineare Transformationen" with subtitles:

Kapitel I: Das gruppentheoretische Fundament der Tensorrechnung,

Kapitel II: Die Darstellungen der Komplexgruppe und der Dehnungsgruppe,

Kapitel III: Struktur der halb-einfachen Gruppen,

Nachtrag (3 pages).

The three papers and the "Nachtrag" are reprinted in "Selecta Hermann Weyl" (Birkhäuser, Basel 1956) on pages 262–366.

Chapter I is wholly devoted to the special linear group $G = SL(n, \mathbb{C})$. First, the irreducible representations of the Lie algebra L_G are determined by Cartan's infinitesimal method. Next, Weyl proves an extremely important theorem:

Every representation of the Lie algebra L_G, and hence every representation of the group G, is completely reducible.

Weyl proves this theorem by means of what he himself calles the "unitary trick", which I now shall explain. If one restricts oneself to *unitary transformations*, which transform the Hermitean form

$$x_1 \bar{x}_1 + \ldots + x_n \bar{x}_n$$

into itself, one obtains a compact Lie group H. On every Lie group one can introduce an invariant volume element dV. Now if one has a representation of a compact Lie group H, one can use an idea of Hurwitz. Starting with any positive Hermitean form in the space of the representation, one applies to it all transformations of H and integrates over H. Thus one obtains a positive Hermitean form invariant under the transformations of the representation. Every invariant subspace defines an orthogonal invariant subspace, and it follows that every representation of H is completely reducible.

If one has an *infinitesimal* representation of H, one can always integrate it, thus obtaining a global representation of a covering manifold of H. Weyl proves that the unitary group $H = SU(n, \mathbb{C})$ is *simply connected*, which implies that every infinitesimal representation can be extended to a global univalued representation of *H*. We have just seen that these global representations are completely reducible, so we may conclude that every infinitesimal representation of *H* is completely reducible.

Next, Weyl proves: *If the infinitesimal transformations of H transform a linear subspace into itself, so do the infinitesimal transformations of G.* Since the representation space of the Lie algebra L_H is a direct sum of irreducible subspaces, the same holds for L_G. This completes the proof of Weyl's theorem.

Weyl's proof is very remarkable. Hurwitz' idea of integration cannot be directly applied to the group *G*, because *G* is not compact. Therefore Weyl first passes from the group *G* to the compact group *H*, next to the Lie algebra L_H, next to L_G, and finally to *G*:

A genius like Weyl was needed to find this proof.

The second paper in Math. Zeitschrift 24 consists of the Chapter II and III. In Chapter II Weyl applies his methods to the symplectic group $Sp(2n, \mathbb{C})$ and to the rotation group $O(n, \mathbb{C})$. In the case of the symplectic group the unitary subgroup *H* is again simply connected, which implies that all infinitesimal representations generate univalued global representations. Once more, all representations are contained in the tensor representations.

In the case of a complex rotation group *G* one first forms the subgroup *H* of *real* rotations. This group is compact, but not simply connected. It has a simply connected two-fold covering group *H**. The Lie algebra of *H** is the same as that of *H*; I shall call it L_H. Now one can apply the "unitary trick", passing from *G* to *H**, from *H** to L_H, and from L_H to L_G. Result: all infinitesimal representations of *G* are completely reducible, and the irreducible representations are those found by Cartan. They generate one-valued or two-valued representations of *G*. The simplest two-valued representations are the "spinor representations".

In Chapter III Weyl presents a new simplified derivation of Cartan's structure theory of semi-simple Lie groups. He shows that the "unitary trick" can be extended to all semi-simple groups.

In Chapter IV Weyl shows that his representation theory can be extended to all semi-simple groups, including the exceptional groups.

In 1927, Weyl published, together with his pupil F. Peter, a paper entitled "Die Vollständigkeit der primitiven Darstellungen einer geschlossenen kontinuierlichen Gruppe", Math. Annalen 97, p. 737–755. In this paper the authors first note that the representations of a compact Lie group *G* are equivalent to unitary representations. Next they prove that the matrix elements $e_{ik}(s)$ of the irreducible representations form a *complete orthogonal* set of functions on the group *G*. The proof is based on the theory of integral equations.

John von Neumann

In the papers of Cartan and Weyl all representations of Lie groups were supposed to satisfy certain conditions of differentiability. In 1929 John von Neumann showed that these conditions can be replaced by very weak continuity conditions. See J. von Neumann: "Über die analytischen Eigenschaften von Gruppen linearer Transformationen und ihrer Darstellungen", Math. Zeitschrift 30, p. 3–42. A part of von Neumann's results had already been presented in an earlier paper: "Zur Theorie der Darstellungen kontinuierlicher Gruppen", Sitzungsber. preuss. Akad. 1927, p. 76–90. Both papers are reproduced in John von Neumann's Collected Works, p. 134–150 and 509–549.

Von Neumann defines the absolute value of a matrix A with complex coefficients as

$$(5) \qquad |A| = \sqrt{\Sigma |a_{\mu\nu}|^2}.$$

He first considers groups of matrices and next their representations. If G is a group of matrices and $\bar{G}$ its closure, he proves that the component of E in $\bar{G}$ is a Lie group H, generated by infinitesimal linear transformations

$$U = a_1 V_1 + \ldots + a_k V_k,$$

whereas $\bar{G}/H$ is a discrete group.

Concerning the representations of G, von Neumann proves: If the variation of a representation $D(A)$ in a neighbourhood of E, defined by means of the metric (5), is less than 2, the representation is continous. In a suitable neighbourhood of E the matrix elements of $D(a)$ are convergent power series in the real and imaginary parts of the matrix elements $a_{\mu\nu}$.

In a paper "Stetigkeitssätze für halbeinfache Liesche Gruppen" (Math. Zeitschrift 36, p. 780–786, 1933) I have given a simpler proof of von Neumann's second theorem, and I have proved that all representations of compact semisimple Lie groups are continuous. This latter result had been obtained already in 1930 by Elie Cartan: "Sur les représentations linéaires des groups clos", Comment. Math. Helv. 2, p. 269–283. It has been considerably generalized by Hans Freudenthal in a paper "Die Topologie der Lieschen Gruppen als algebraisches Phänomen", Annals of Math. 42, p. 1051–1074 (1941).

In 1934, John von Neumann published an extremely important paper "Almost Periodic Functions in a Group I", Transactions Amer. Math. Soc. 36, p. 445–492. The introduction to the paper begins thus:

The object of the present paper is to extend H. Bohr's famous theory of almost periodic functions to arbitrary groups, and to show that it gives just the maximum range over which the fundamental results of Frobenius-Schur representation theory and its extensions by Peter and Weyl hold. We shall see in particular that all bounded linear representations of a group are equivalent to unitary representations and belong to this class. Another point of importance is that we free ourselves completely from all topological assumptions (such as continuity, etc.) by the use of a definition of almost periodicity due to Bochner. Thus we find that the general theory, which applies to every group G whatsoever, is completely free from topological assumptions, but all of its results (for example, all series expansions) have a property of closure; if applied to functions which are continuous in a certain topology, they will lead only to functions of the same kind.

If f and g are complex-valued functions on a group G, their distance $D(f, g)$ is defined as the least upper bound of the difference $|f(x) - g(x)|$:

$$D(f, g) = \text{l.u.b.} \, |f(x) - g(x)|.$$

A set M of such functions is called *conditionally compact* (c.c.), if every sequence $f_1, f_2, \ldots$ contains a subsequence such that

$$D(f_\mu - f_\nu) \to 0 \quad \text{as} \quad \mu, \nu \to \infty.$$

A function $f(x)$ in G is called *almost periodic* if the set R_f of all functions $f(xa)$ and the set L_f of all functions $f(ax)$ are both conditionally compact. This definition is due to S. Bochner (Math. Annalen 96, p. 119–147, 1927). If G is the additive group of real numbers and if f is continuous, Bochner's definition is equivalent to that of Harald Bohr.

The *convex* $Co(M)$ of a set M of functions is defined as the set of all linear combinations

$$\alpha_1 f_1 + \ldots + \alpha_n f_n, \; \alpha_i \geq 0, \; \Sigma \alpha_i = 1, \quad f_i \in M.$$

If f is almost periodic, there exist a constant towards which a subsequence of $Co(R_f)$ converges uniformly. Such a constant is called a *right mean* of the function f. If $f(x)$ is almost periodic, it has exactly one right mean, which is also a left mean, and which is called the *mean*

$$Mf = M_x f(x).$$

It is easy to see that the matrix elements $d_{ik}(x)$ of a bounded representation of G are almost periodic, Now, if one starts with a positive Hermitean form

$$h = \bar{x}_1 x_1 + \ldots + \bar{x}_n x_n$$

and if one applies to this form all transformations of a bounded representation and forms the mean Mh, it follows that every bounded representation of G is unitary and completely reducible.

If $d_{ij}(x)$ are the matrix elements of an irreducible bounded representation, the orthogonality relations of Issai Schur can be generalized as follows:

(6) $$M_y d_{ij}(xy^{-1}) d_{kl}(y) = (1/n) \, \delta_{jk} \, d_{il}(x)$$

(7) $$M_y d_{ij}(xy^{-1}) d'_{kl}(y) = 0, \quad \text{if } D' \text{ is in equivalent to } D.$$

If $f(x)$ and $g(x)$ are almost periodic, the cross product $f \times g$ is defined as

$$h(x) = M_y [f(xy^{-1}) g(y)].$$

Under this multiplication, the almost periodic functions form a ring R. If the group G is finite, this ring is just the group algebra.

In the ring R a *scalar product* (f, g) can be defined as

$$(f, g) = M_y[f(y)\overline{g(y)}]$$

and a *norm* $N(f)$ as

$$N(f) = \sqrt{(f, f)}.$$

A set of functions f_i is called *complete* if the linear combinations

(8) $$\gamma_1 f_1 + \ldots + \gamma_r f_r$$

are everywhere dense in the vector space R, that is, if every almost periodic function f can be approximated by a linear combination (8):

$$N(\gamma_1 f_1 + \ldots + \gamma_r f_r - f) < \varepsilon.$$

In order to prove the completeness of the $d_{ik}(x)$, John von Neumann considers the integral equation

$$f \times f^\dagger \times \psi = \lambda \psi$$

with in which $f^\dagger(x)$ is complex conjugate to $f(x^{-1})$.

This method is a generalization of a method employed by F. Peter and H. Weyl in Math. Annalen 97 (1927), p. 737–755. In §14 of my book "Gruppen von linearen Transformationen" I have proved the completeness of the $d_{ik}(x)$ by another method due to Gottfried Köthe. In my proof I have used the theory of generalized Hilbert spaces developed by Franz Rellich in Math. Annalen 110, p. 342–356 (1934).

Von Neumann proved a little more than the completeness of the functions $d_{ik}(x)$. He proved: Every almost periodic function $f(x)$ can be *uniformly* approximated by a linear combination of the $d_{ik}(x)$.

All theorems just mentioned are also true if one restricts oneself to continuous bounded functions $f(x)$ on a topological group G and to continuous bounded representations.

A topological group is called *maximally almost periodical*, if for any two group elements a and b an almost periodic function f exists such that

$$f(a) \neq f(b).$$

Examples of such groups are the compact topological groups and the abelian locally compact topological groups.

In 1935, Hans Freudenthal proved that there are no other maximally almost periodical groups than those just mentioned and their direct products. More precisely:

Every connected, locally compact, separable, maximally almost periodical topological group is a direct product of the translation group of a Euclidean n-dimensional space $\mathbb{R}^n$ and a connected compact topological group.

See H. Freudenthal: Topologische Gruppen mit genügend vielen fastperiodischen Funktionen, Annals of Math. 37, p. 57–77 (1935).

Index

Abel, Niels Henrik 76, 85–88, 109, 117, 149
Abelian functions 124–125
– groups 88, 149–150
– integrals 133
abscissa 71
absolute value 216, 261
absolutely irreducible 249
Abu al-Jud 29
Abu Ǧafar 41
Abu Yusuf 16
accessory irrationalities 156
accompanying system 206
Ackermann, M. 160
adjoint group 168
adjunction 86, 109
Aeberli, G. 188
Albert, A.A. 213, 215
d'Alembert 95
Alexander, J.F. 222
Alfonso d'Avalos 55
algebras 175-217, 237-251
al-jabr see jabr
al-Khwārizmī see Khwārizmī
almost periodic functions 222, 261-263
alternating group 122
amicable numbers 21–23
Anderson, Alexander 65, 67
Annibale della Nave 59
Antanairesis 30
Apollonios 6, 13, 70, 71, 74
Archimedes 6
Archytas 39
area of a circle 5–6, 65
– – – segment 7
Argand, Jean Robert 178
Arithmetic of quaternions 186-188
Arrighi, Gino 49
Ars magna 53–59
Artin, Emil 210–211, 214
Aryabhata 6, 13
Aryabhatiya 6, 13
Auerbach, H. 158
Ausdehnungslehre 191
auxiliary equation 137–138
Azra, J.-P. 103

Bachet, C.G. 186
Baer, Reinhold 211

Bancroft, George 203
Banu Musa 39
Barhebraeus 17
Becker, Oscar 30
Benedetto of Florence 42–44
Bernoulli, Johann 17
Betti, Enrico 114–115, 117, 124
Biaggio 43
Bieberbach, L. 158
Biquaternions 188–189
Biruni 10
Blichfeldt 158, 159
Bochner, S. 222, 261–262
Bohr, Harald 222, 261–262
Bollinger, Maja 114
Bolyai 143
Bolza, Oskar 238
Bombelli, Rafael 59–61, 147, 177
Boncompagni, Baldassare 9, 35
Borchert 139
Bortolotti, E. 53
Bourgne, R. 103
Brahmagupta 6, 10, 14
Brahmasphutasiddhanta 6, 10, 14
Brandt, H. 199
Brauer, Richard 194–197, 213–217, 257
– class 213, 215
– group 213, 215
Bravais 119
Brioschi 117, 133
Burckhardt, J.J. 10, 12, 120
Burkhardt, Heinrich 84, 139
Burnside, William 237-238

Cajori, F. 34
Cardano, Gerolamo 52–60, 76, 79, 177
Carmody, C.F. 17
Carnot 84
Cartan, Elie 165–174, 196, 204, 208–210, 238,
    252, 254–261
Cartan, Henri 258
Casorati 156
Casus irreducibilis 56, 60, 67–68
Cauchy, Augustin Louis 76, 84–86, 88, 103,
    109, 121, 137–138, 178, 192
Cayley, Arthur 117, 126, 137, 141–144, 147,
    150–152, 154, 183–185, 190, 227
Cayley-Klein parameters 186

Cayley numbers 183
censo 42
census 38
centre 140, 246, 249
character 209
characteristic 110
– equation 168, 208
– function 189
– roots 168–171, 208, 229
characters of abelian groups 218–223, 250
– of finite groups 223–236
– of quadratic forms 218–220
– of topological groups 222
Chevalier, Auguste 105, 107
Chevalley, Claude 183, 197
Chiu Chang 34
Chwolson, D. 15–16
circumference of circle 5–7, 14
Clagett, M. 39
Clark, W.E. 6
class of quadratic forms 219
Clavius 69
Clebsch, Alfred 117, 123, 125–126
Clifford, William Kingdon 188–189, 192
– algebras 192–196
– numbers 194–196
cogredient variables 256
Cohn, P.M. 162
Colebrooke 6, 14
commentaire sur Galois 121
complete matrix algebra 189, 207
– set of functions 263
completely reducible 239, 246, 248, 259
complex 210
– numbers 56, 60–61, 95, 177–178
composition of quadratic forms 148–149, 220
conditionally compact 262
congruences 121
congruent modulo H 122
congruentus 40–41
congruum 40–41
Conica 71, 75
conic sections 72, 75, 141–144
conjugate subgroups 114
continued fractions 103
convex 262
cosa 44
coset 108–109, 138, 148
Cotes, Roger 177
covector 172
Coxeter, H.S.M. 173
crossed product 199–200
cubic curve 125
– surface 126
cubo 42
curve, cubic 125
– quartic 126, 128–131
cyclic algebra 200–201, 215, 217

cyclotomic equation 77, 80, 89–94

Dantzig, David van 216
Dardi 32, 47–52
decimal fractions 68–69
– point 69
Dedekind, Richard 157, 204–205, 220–221, 223–228, 233, 250
Degen, C.P. 184
denarii 50
Descartes, René 21, 64, 69, 72–75, 177
determinant 189, 218, 250
Deuring, Max 211, 213–216
Dickson, Leonard Eugen 123, 186, 200–201, 209, 214
Dieudonné, Jean 118, 121, 123, 134
difference algebra 210
differential equations 146–147
Diophantos 13
Dirac, P.A.M. 194–196
direct product 212, 244
– sum 206, 209–210, 245–249
Dirichlet, P.G. Lejeune 220, 222, 235
discriminant 99, 148, 250
distance, Euclidean 142–143
– non-Euclidean 141–144
division algebras 213–217
Dold, Yvonne 29
double module 246
– tangents 128
doubling the cube 26, 65
Drower, E.S. 15
duel of Galois 104–105
Dupuy, Paul 103
Dyck, Walter von 137, 147, 151–154
Dynkin, E. 171

Edwards, H.M. 107
Eichler, Martin 199
elementary divisors 122–123
ellipse 71-72
Engel, Friedrich 147, 161–164, 166, 169
epicycle 11
equant point 13
equation of anomaly 11
– of centre 11
equations, biquadratic 43, 45, 47–52, 78
– cubic 25–29, 34, 45–46, 47–50, 52–61, 66–67, 78–80, 82
– Diophantine 40, 198
– quadratic 7–9, 18–21, 25, 38, 42–43, 64, 74
– quintic 78, 81–83, 86–88, 109, 133
Era Yazdigerd 9, 11
Erlanger Programm 144
Escott, Edward 23
Euclid 6, 8–9, 19–20, 30, 36, 39, 74, 89–90
Euler, Leonhard 23, 78, 82, 89, 95, 103, 120, 140, 147–148, 177–178, 184, 186

Eutokios 26
exterior products 192

factor group 121–122, 156
Fazari 10
Fedorow, E.S. von 120
Fermat, Pierre 21, 69–72
Ferrari, Lodovico 56–59, 76
Fibonacci 32–33
–, series of 36
fields 109
–, finite skew 214
–, P-adic 214–217
Fihrist 16
Finzi, Mordechai 47
Fiore, Antonio Maria 54
Fitzgerald, E. 24
Flos 33–34, 40
Fontana, Niccolò 55
Fontenex 95
Fourier 103
– series 156, 222
Fowler, D.H. 30
fractional linear transformations 124, 139,
    145–146, 153
Franci, Raffaela 41–44, 45, 82
Frederick II 34, 40
Freudenthal, Hans 146, 165, 171, 174, 261,
    263
Frobenius, Georg 139, 140, 155, 158, 190, 214,
    223–236, 237–238, 242
Frucht, R. 243
full matrix algebra 189, 207, 209, 212–213,
    217
functions, abelian 124–125
– elliptic 131, 133
– symmetric 76–77
– transcendental 131–133
fundamental domain 153
– equation 190
– theorem of algebra 94–102
– – on abelian groups 149–150, 155
– theorems on Lie groups 163–164

Galois, Evariste 76, 85–86, 103–118, 124, 147
– fields 109–111, 121, 147, 214
– group 107, 120, 131
– theory 77, 80, 103, 105–109, 117–118, 124,
    156
Gammafunction 156
Gantmakher, F. 174
Ganz, Solomon 5–7, 14–16
Gauss, Carl Ludwig 76, 78–79, 81, 89–102,
    105, 109, 147–149, 178, 204, 218–221
Geiser, M. 130
general equation 137
– linear group 123
generic 125, 170, 202

genus 219
geography 9
geometry, descriptive 143
– elliptic 143
– projective 143
– hyperbolic 143
Gerard of Cremona 32, 39
Gierster 124
Giovanni da Palermo 40
Girard, Albert 177
global method 253–254
Gödel, Kurt 157
Goldbach 184
Gordon, W. 194
Goursat, F. 159
Grace, J.H. 242
Grassmann, Hermann Günther 191–192
Graves, John 179–181, 183–184
Gray, J.J. 120
Griffith, I.W. 199
Gross, Herbert 188
group algebra 190–191, 227, 251
– characters 218–236, 239
group, cyclic 119, 146
– determinant 223–236
–, dihedral 119, 146, 225
–, icosahedral 119, 146, 234
–, octahedral 119, 146, 234
– of an equation 107
– pair 221
– table 151
–, tetrahedral 119, 146, 234
groupe abélien 124
– hypoabélien 124
groups 115–116, 121–124, 137–154
–, abstract 137, 147–154
–, continuous 160–174
–, finite linear 119, 146, 158–159
– of motions 118–120
– of permutations 84, 107, 115–116, 121, 133–
    134, 137–140
– of transformations 160–174
– with operators 244

Haar, A. 221
Haji Khalfa 4
Hajjaj 14–15, 20
Hamilton, Archibald 179
Hamilton, William Rowan 147, 152, 178–185
Hankel, Hermann 14, 191–192
Harun al-Rashid 3
Hasse, Helmut 199–200, 214–217
Hauptgruppe 144
Hawkins, Thomas 122, 207, 223–227, 234,
    237–238
Hazlett, Olive 214
Heegard, P. 114
height 253–254

Henderson, A. 126
Henri III 63
Henri IV 63, 65
Hensel, Kurt 214
Hermann, R. 160
Hermite, Charles 112–113, 117, 123, 124, 131, 133
Hermitean form 158, 239–240
Heron 6–7
Hesse, Ludwig Otto 117, 125
Hindu numerals 9, 14, 33, 35
Hippokrates 26
Hölder, Otto 121, 155–157
holoedric 122
homomorphisms 122, 230–231
homomorphy theorem 245
Horner, W.G. 34
House of Wisdom 3, 14
Hurwitz, Adolf 158, 185–188, 259–260
hyperbola– 28–29, 71–72
hypercomplex numbers 177–217, 225

Iamblichos 15
Ibn al-Qifti 10, 17
ideal 167
–, left 211
–, right 211
–, two-sided 209–210
idempotent 203–204, 245
imaginary numbers 56, 61, 177
indecomposable 248
indéterminée 80
index 121, 138
infinitesimal method 254, 259
– transformation 162–164
inflexion points 125
inheritance 5, 7
integrable 166
invariant subgroup 108, 167
irreducible group 133
– polynomial 106
– representations 230, 236, 237, 239–240, 242, 243, 248–251
isomorphism 122
isomorphy theorems 242

jabr 4–5, 64
Jacobi, Karl Gustav Jakob 112, 162, 186–187
– identity 162
– symbol 162
Jacobson, Nathan 211–212
Jamshid al-Kashi 34
Jewish calendar 7
John of Palermo 34, 40
Jordan, Camille 85, 117–134, 139–141, 144, 147–148, 159
Jordan-Hölder Theorem 121, 156, 244, 247
Jordan normal form 122, 189
Jordanus de Nemore 39
Juschkewitsch, A.P. 29–31

Kahn, D. 63
Kampen, E.R. van 222–223
Kankah 11
Karaji 4
Kennedy, E.S. 10–11
Khorezm 3
Khwārizmī 3–15, 20, 38, 48
Kiernan, B.M. 84, 112, 114
Killing, Wilhelm 166–170, 207
Kimberling, Clark 225
Kirkman 124
Klein, Felix 118, 143–146, 152, 158, 185, 196, 238
Klein, Oscar 194
Kollros, L. 103
Koran 15–16
Köthe, Gottfried 211, 213, 263
Kronecker, Leopold 88, 113, 117, 123, 133, 137, 147, 149–151, 154–155
Krull, Wolfgang 244
Kummer, E.E. 117, 126

Lacroix 104
Lagzange, Joseph-Louis 76, 78–81, 83, 85, 86, 88–90, 95, 103, 109, 137–138, 148, 186, 198
– resolvent 78, 80, 109
Laguerre, E. 190
Laplace equation 155
Latham, Marcia 72
Latimer, C.G. 199
Lebesgue, Henri 77, 117
legacies 5, 7
Legendre, Adrien Marie 23, 84, 184, 198
– symbol 219
Leibniz 69
Leonardo da Pisa 32–42, 48
Levitzki, Jakob 211
Liber abbaci 33, 35–39
– embadorum 39
– quadratorum 33
Libri, Guillaume 47
Lie algebras 160–174, 254–260
– groups 146, 158, 160–174, 252–263
– product 62, 203
Lie, Sophus 118, 144–145, 160–174, 196, 205, 252
Lie's theory 160–174
linear transformations 122–124, 158–159, 237–263
Liouville 103, 112–113
Lipschitz, R. 193
Littlewood, D.E. 242, 253
Liu Hui 6
Lobatchewsky 143
local-global principle 199, 216–217
locally isomorphic 163
loci, plane 69–71
– solid 69–71

Loewy, A. 158, 239
Lorentz group 195
– transformation 195
Luca Pacioli 32, 46–47
Luckey, P. 18, 34

MacDuffee, C.C. 189
Mahoney, Michael 69
Maillet 139
main theorem on abelian groups 149–150, 155
– – on division algebras 217
Malfatti, Gianfrancesco 76, 81–83
Ma'mun 3–5, 9, 14
Mandaeans 15
Manfredi 82
Mansur 10
Marre, Aristide 6
Maschke, Heinrich 238–240
Maslama al-Majriti 9
Mathieu, E. 121, 123–124, 139
matrix 122, 189–190
Maurits, prince 68
maximally almost periodic 263
Muhammad ben Musa 3–15
– – – ben Sakir 16–17
multiplication 153
Mazzinghi, Antonio 44
mean 222, 262
Menaichmos 26
mensuration 5–7
meroedric 122
Mersenne 70
Method of the Indians 12
– – – Persians 12
Miller, G.A. 139
Mishnat ha-Middot 5–7
Mitchell, H.H. 159
modular equation 123
– functions 133
– ideal 212
moduli, law of 180–184
Moivre, Abraham de 89, 17
Molien, Theodor 190, 206–209, 234, 237–238, 240, 250
monodromy group 113, 131–132
Montgomery, L. 165
Moore, E.H. 158, 238, 239
muqabala 4–5

Nadim 16
Napier 69
Nasir ad-Din 31
Needham, J. 34
negative numbers 46
Nehemiah 6
neighbourhood 161, 252
Netto, E. 95, 139
Neuenschwander, E. 156

Neugebauer, Otto 9, 10, 17, 71
Neumann, John von 222–223, 242, 261–263
Newton, Isaac 69
Nikomachos 16
Nilideal 211
nilpotent 203, 209
– ideals 246
Noether, Emmy 199, 211, 213, 215–217, 244–251
non-Euclidean geometry 141–144
non-quaternion systems 205
norm-residue symbol 216
normal divisor 108
– form 122, 189
– simple algebra 212–213
– subgroup 167
North, John D. 141

obliquity 17
O'Brien 192
octonions 183–184
Omar Khayyam 3, 24–32, 64
ordinate 71
Ore, Oystein 55
orthogonal groups 123, 159, 166, 196–197, 257
orthogonality relations 222, 228, 262
Ostrowski, Alexander 97, 216

Page, M. 166
Pappos 63, 74
parabola 25–28, 71–72
Pasch, Moritz 190
Peirce, Benjamin 203–204
Peirce, C.S. 203
– decomposition 203–204
periods 92–94
permutations 85, 107
Persians 10
Peter, F. 260, 261
Petersen, M.J. 140
PGL 123, 166, 252–253
Philo of Byzantium 39
Picard, Emile 103
Piero della Francesca 49
Plato 39
– of Tivoli 39
PO 252
Poincaré, Henri 114–115, 203, 205
Poisson 104–105, 107
polar form 207
– property 207
polygons, regular 89–91
Pontryagin, L. 221–223
positive Hermitean form 158
– quadratic form 172, 219
power sums 77
Practica geometriae 39

precession 17
primitive algebra 207
– element 111
– form 148, 153, 218
– group 116, 121, 133
– root of unity 78, 80
projective group 123, 139, 166
Proklos 16
proper decomposition 108
pseudo-nul 209
PSL 123–124
PSp 123, 133, 166, 252
Ptolemy 9, 10, 13, 17
Puiseux, Victor 112–113, 131
Pythagoras, Theorem of 6, 37
Pythagorean metric 257–258
Pythagoreans 16, 89

quantics 141
quartic curve 128–131
quaternion group 225, 243
– systems 205
quaternions 179–188
–, generalized 197–199
–, primary 187
Qutrubull 3

radical 210–212, 248–249
Rashed, Roshdi 24
rational 109
real numbers 69, 95, 97, 157
reducible 106, 206
Rellich, Franz 263
representation group 243
– module 246
–, reducible 247
–, regular 202, 229, 237
representations of algebras 190, 230, 237–251
– of groups 190, 237–251
Richard, Louis 112
Richardson, A.R. 242, 253
Riemann, Bernhard 114
Roberval 70
Rodrigues, Olinde 120, 140, 186
Rohr, H. 206
Romanus, Adrianus 65
roots of unity 78, 80, 82, 108
Rosen 4, 7
rotations 118–120, 185–186, 193–194, 258
Ruffini, Paolo 34, 76, 83–84, 88, 137

Sabians 15–16
Sahili, Aydin 24
Saint-Venant 192
Saliba 4
Salmon 117, 126
scalar product 263
Scheffers, Georg 205

Schering, Ernst 149, 150
Schmidt, Otto 244
Schneider, Ivo 177
Schoenflies, A. 120
Schreier, Otto 157, 161
Schur, F. 165
Schur, Issai 239–240, 243, 253–254, 262
Schwarz, Hermann Amandus 204
Scipione del Ferro 52–54, 59, 76
Scotus, Michael 34
Séguier, J.A. de 123
semi-simple algebras 209
– – Lie groups 170–171
– – rings 212
series of composition 121, 156, 244
– – Fibronacci 36
Serret, Joseph-Alfred 113–114, 117–118, 121, 124, 138
similar functions 80
– substitutions 85
simple algebra 207
– Lie algebra 172–174
– group 166, 170–171
– module 212
– ring 211
Sindhind 10–11
skew fields 214
SL 123, 171, 255–257, 259
Smith, David Eugene 72
Sohnke, Leonhard 120
solvable 109, 115, 124, 149, 166, 169, 205
Sp (2n, p) 123
space problem 166–167, 259
Speiser, Andreas 158
spinning electron 194
spinors 195–197, 257
splitting field 213
square root 37, 73
Steiner, Jakob 117, 126
Stevin, Simon 68–69, 74
Stickelberger 155
structure of algebras 190, 202–217
Study, Eduard 183, 185, 189, 190, 204
Stufe 253
substitutions 85, 107
sums of squares 184–186
Suter, Heinrich 9, 11, 16, 17
Sylow, M.L. 124, 139–140, 155
symmetric functions 76–77
– group 137–139, 224, 234, 240–243
system determinant 250

Tabari 3
Taber 190
Tabit ben Qura 3, 15–23
table of Sines 9

tableau 241–242
tables, astronomical 9–13
Tartaglia 52, 54–59, 76
Taton, René 104
tensor representation 254
Theodorus 33, 40
Threlfall 159
Toomer 3, 9, 10, 13
topological groups 221–223, 261–263
Toti Rigatelli, Laura 42–44, 82
trace 189, 236, 250
transformations, fractional linear 124, 139, 145–146
– linear 122–124, 144–145
transitive 115–116, 121
trepidation 17
triangle, rectangular 6
trisection 65, 67
Tropfke, Johannes 35–36, 68–69

Umlauf, Arthur 168
unitary 259–260

Valentiner, H. 159
valuations 216
–, archimedic 216
–, P-adic 216
Vandermonde 76–79, 80
Van Egmond, Warren 32–33, 47
Venkov, B. 199
Viète, François 63–68, 76
Vogel, Kurt 9, 34–39, 42

Waerden, B.L. van der 171, 223, 261
Wahlin, C.E. 199
Waring, Edward 76–77
Warren, John 178, 180
wealth 4, 19–20
Weber, Heinrich 77, 130, 137, 147, 151–154, 221
Wedderburn, J.H. Maclagan 197, 201, 209–210, 214
Weierstrass, Karl 122–123, 156, 204, 235
weight 255
Weil, André 69, 198
Wertheim, G. 114
Wessel, Caspar 178
Weyl, Hermann 171–173, 194–197, 242, 253–254, 257–261
Weyl's group 172
Weyr 190
Wiedemann, H. 14
Witmer, T.R. 56
Witt, E. 214
Woepcke, F. 21, 24, 29, 42
Wussing, H. 137, 147

Yaqub ben Tariq 11
Yazdigerd 11
Young, A. 240–242
Youshkevitch, A.P. 29–31

zij 10
Zij-i Shah 10
Zippin, L. 165

B. L. van der Waerden

Geometry and Algebra in Ancient Civilizations

1983. 98 figures. XII, 223 pages
ISBN 3-540-12159-5

Contents: Pythagorean Triangles. – Chinese and Babylonian Mathematics. – Greek Algebra. – Diophantos and his Predecessors. – Diophantine Equations. – Popular Mathematics. – Liu Hui and Aryabhata. – Index.

From the reviews:

"The chief aim of this fascinating book is to demonstrate that mathematical ideas were communicated between civilizations in Europe, Asia and Africa earlier than is commonly supposed. As preparation for this investigation, the author carefully describes radiocarbon dating, which was introduced in 1946 by W. F. Libby and corrected by later authors using the tree-ring method. Ancient skills can be observed in megalithic monuments built between 4800 and 2000 B.C. in Portugal, Spain, Malta, Brittany, Ireland, Scotland and England. The enormous stones in 'henges' such as Stonehenge and Woodhenge are seen to be located on circles, ellipses and other ovals, probably constructed by stretching a closed rope round one, two or three pegs. The required positions of such sets of three pegs indicate knowledge of Pythagorean traingles $(3, 4, 5)$, $(5, 12, 13)$ and $(12, 35, 37)$ in the Neolithic Age. The author cites several reasons for this belief that the geometric and astronomical skill necessary for such architecture must have developed in just one center and spread to both Egypt and western Europe.

Careful comparison of ancient texts seems to indicate a tradition of teaching mathematics by means of well-chosen sequences of problems and solutions, a tradition which seems to have originated somewhere in neolithic Europe and spread towards Greece, Babylon, India and China. Page 45 lists fifteen examples of striking similarities between various ancient civilizations. In particular, the author shows that Babylonian and Chinese algebra must have had a common source.

… This is a book which every teacher of mathematics should possess, not only for its revelation of unexpected historical connection but also for its wealth of worked examples. Surely it must be stimulating for students to see how these tricky problems were tackled thousands of years ago, before the invention of trigonometry."

College Mathematical Journal

Springer-Verlag
Berlin
Heidelberg
New York
Tokyo

L. Euler

Elements of Algebra

Translated from the French by J. Hewlett
With an Introduction by C. Truesdell
Reprint of the 5th edition Longman, Orme, and Co.,
London 1840

1984. LX, 593 pages
ISBN 3-540-96014-7

Contents: Part I: Containing the Analysis of Determinate
Quantities: Of the Different Methods of Calculating Simple
Quantities. Of the Different Methods of Calculating
Compound Quantities. Of Ratios and Proportions. Of Algebra-
ic Equations, and of the Resolution of those Equations. Part II:
Containing the Analysis of Indeterminate Quantities. – Addi-
tions by M. La Grange.

B. Chandler, W. Magnus

The History of Combinatorial Group Theory: A Case Study in the History of Ideas

1982. 1 figure. VIII, 234 pages
(Studies in the History of Mathematics and Physical Sciences,
Volume 9)
ISBN 3-540-90749-1

Contents: The Beginning of Combinatorial Group Theory. –
The Emergence of Combinatorial Group Theory as an Inde-
pendent Field. – Bibliography. – Index of Names. – Index of
Subjects.

History of Combinatorial Group Theory is a unique account of
the origins and developments of an important and well-defined
field of mathematics.
Combinatorial group theory is a branch of mathematics whose
origins date back a mere one hundred years. One of its foremost contributors, Wilhelm Magnus, is the senior author of
this book. Together with Bruce Chandler he has not only
compiled most of the published literature available on the
subject, but has also integrated material transmitted in oral
communications. In this way, the authors present the field as it
has emerged from a variety of motivations and sources.
Conversely, the applications of combinatorial group theory to
other fields are also covered.

Springer-Verlag
Berlin
Heidelberg
New York
Tokyo